Contents

Antenna building guide

Blueprints and explanations for HAM and DIY antenna builders for all Bands and uses

Intro

This book is intended for Radio Amateurs and others who want to build an antenna. Its explanations are detailed and simple so that everyone can use it: experienced HAM's can find it as a useful resource, beginners can use it as an entrance into the world of antennas or just to build an antenna they need.

The book contains:

- Over 130 antenna designs,
- Blueprints and schematics,
- All data needed for design,
- Computer simulations and Radiation patterns,
- Detailed explanations how to build the antenna,
- Simple explanations how the antenna works,
- Only simple mathematical equations.

What is covered:

- All the most popular and most practical antennas,
- Antennas for all the popular Bands and uses,
- Matching Circuits and BalUns,
- All the popular connectors,
- Cables and their properties

Bands

This book contains blueprints for the next bands and uses:

- Wi-Fi – Wireless Network
  - 2.4GHz
  - 5.8GHz
- Radio broadcast
  - FM (88-108 MHz)
  - AM (535-1705 kHz)
  - TV (200-800 MHz) & DTV
- Amateur Radio - HAM
  - MF
    - 160m (1800-2000 kHz)
  - HF
    - 80m (3500-4000 kHz)
    - 60m (5MHz)

- 40m (7.0-7.3 MHz)
- 30m (10.10-10.15 MHz)
- 20m (14.00-14.35 MHz)
- 17m (18.068-18.168 MHz)
- 15m (21.00-21.45 MHz)
- 12m (24.89-24.99 MHz)
- CB - 11m (27MHz)
- 10m (28.0-29.7 MHz)
  - VHF
    - 6m (50-54 MHz)
    - 2m (144-148 MHz)
  - UHF
    - 70cm (420-450 MHz)
    - 23cm (1240-1300 MHz)
  - Multi and Dual Band antennas
- Cell Phone GSM
  - 850MHz
  - 900MHz
  - 1800MHz
  - 1900MHz

About the Author

The Author has Masters degree in *Experimental Physics* and is currently on postgraduate studies in *Applied Physics* at the *University of Belgrade*. He has many years experience in building, testing and designing antennas.

Copyright information

All text and images are under Copyright and the owner is Aleksandar Ciric.

The exceptions are several images that are in public domain or their license grants reuse with modifications. Those images are free to redistribute or modify.

Universal antenna types

Universal antenna types are the ones which design is being used for many bands (all antennas with different proportions can be used for all the bands, but there is the question of practicality). They are well tested and mostly quite old designs, well before computer simulation. However, they are still widely in use today, and that trend will continue.

In this section we will explore simple dipole antenna and its variations, monopole, Yagi, Biquad, 3D corner and a bit more exotic HB9CV antenna.

Dipole, Inverted V, Delta Loop

Dipole, Inverted V and Delta Loop are used for many bands with one variable – the length of the conductor.

There are only a few theoretical things we need to know about dipoles. The total length of the antenna should be exactly half wavelength[1], with some corrections (we are looking for the term *electrical length*)[2]. That length is to be reduced when building practical antenna. If pipe is being used, or multi-threaded conductor, or angle between radials is changed, all those things affect wave propagation and we need to further reduce the length of radials.

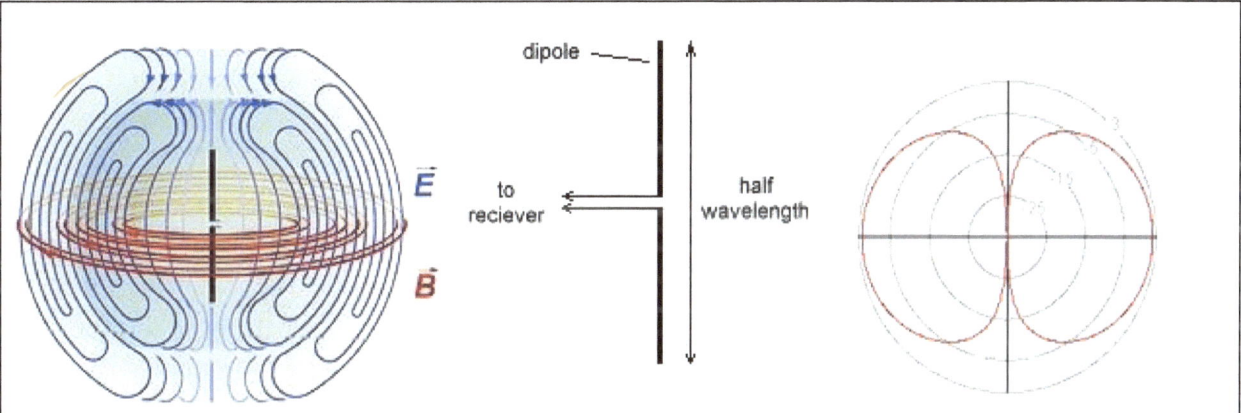

Left image shows propagation of electric and magnetic fields radiated by dipole antenna; Middle image shows the simple schematics of the half-wave dipole; On the right is radiation pattern in dBi.

If the dipole is oriented vertically like on images above, it will radiate what is called vertically polarized EM waves. If it is horizontally oriented it will radiate horizontally polarized waves.

---

[1] There are other types of dipole, but here they are of no interest to us at the moment.
[2] More about electrical length can be seen on Wikipedia: https://en.wikipedia.org/wiki/Electrical_length

If the angle between radials is not 180° like on image, but we shift it to 90°, impedance will be 50Ω instead of 72Ω and we call that antenna Inverted V.

If we make folded dipole, impedance will be 300Ω which is a good match for Twin Lead cable and some devices. We will leave folded dipoles for later.

Half-wave dipole has gain 2.15dBi, Inverted V 1dBd (in practice it is 1.9dBi), Delta Loop 2.78dBi.

Construction
Dipole
Any dipole consists of two radiating elements, insulated ad the feed point. Regular dipole, with the angle of 180°, has feed point impedance of 72Ω. It can be made with a wire or with rods or pipes.

Half-ware dipole is the one most commonly used. The entire length of the antenna is equal to the one half of the wavelength.

Typical equation for dipole: $l = 143/f$, f – frequency in MHz, l – length or the entire antenna in m.

The exact equation is as follows: $l = \frac{1}{2}k\lambda$, where $\lambda = \frac{c}{f}$. k is the adjustment factor, for thin wires (compared to the wavelength) it is equal to 0.98. If conductor is thick (about 1/100 of a wavelength) the adjustment factor is: $k \approx 0.94$.

Inverted V antenna
The construction is the same as with dipole, but the radials are at angle between 90° and 120°. Since Inverted V is being used for low frequencies, it is reasonable to assume that the wire would be used as radials. We must therefore have a mast to which two insulators are attached. On those insulators we attach the wires, like on the image. It is desirable that the feed line be as close as possible, i.e. that the wires' ends are as close as possible.

As with any antenna, it will give the best results if placed high above the ground.

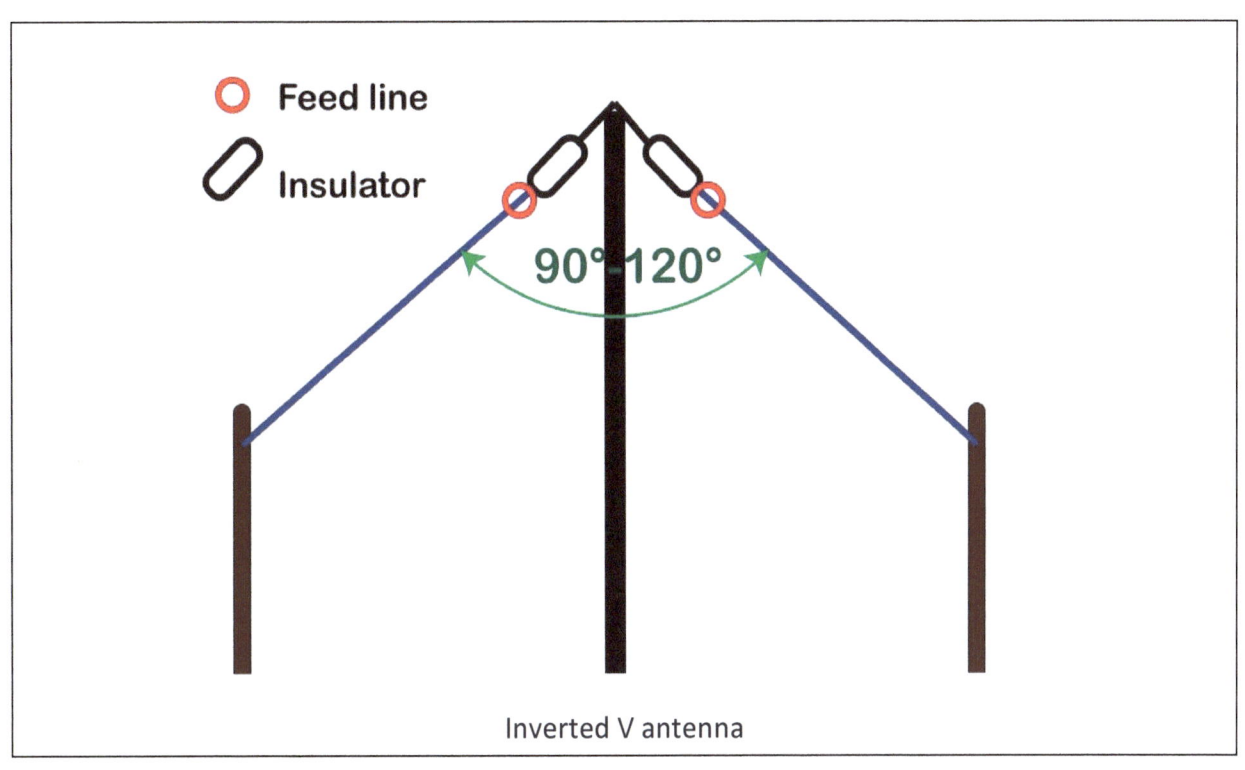

Inverted V antenna

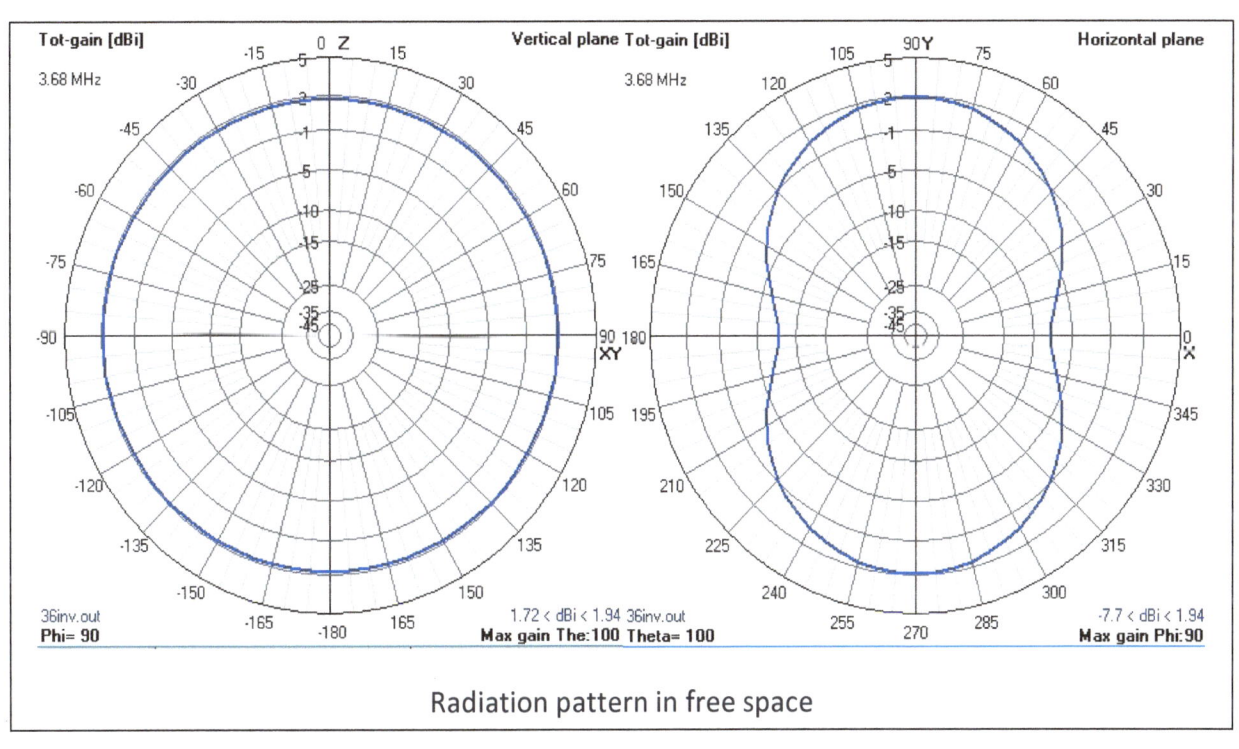

Radiation pattern in free space

Delta Loop

Delta loop is the wire triangle with equal angles of 60° and a feed point at one corner. The measurements are different than for the dipole, and are given in the table below.

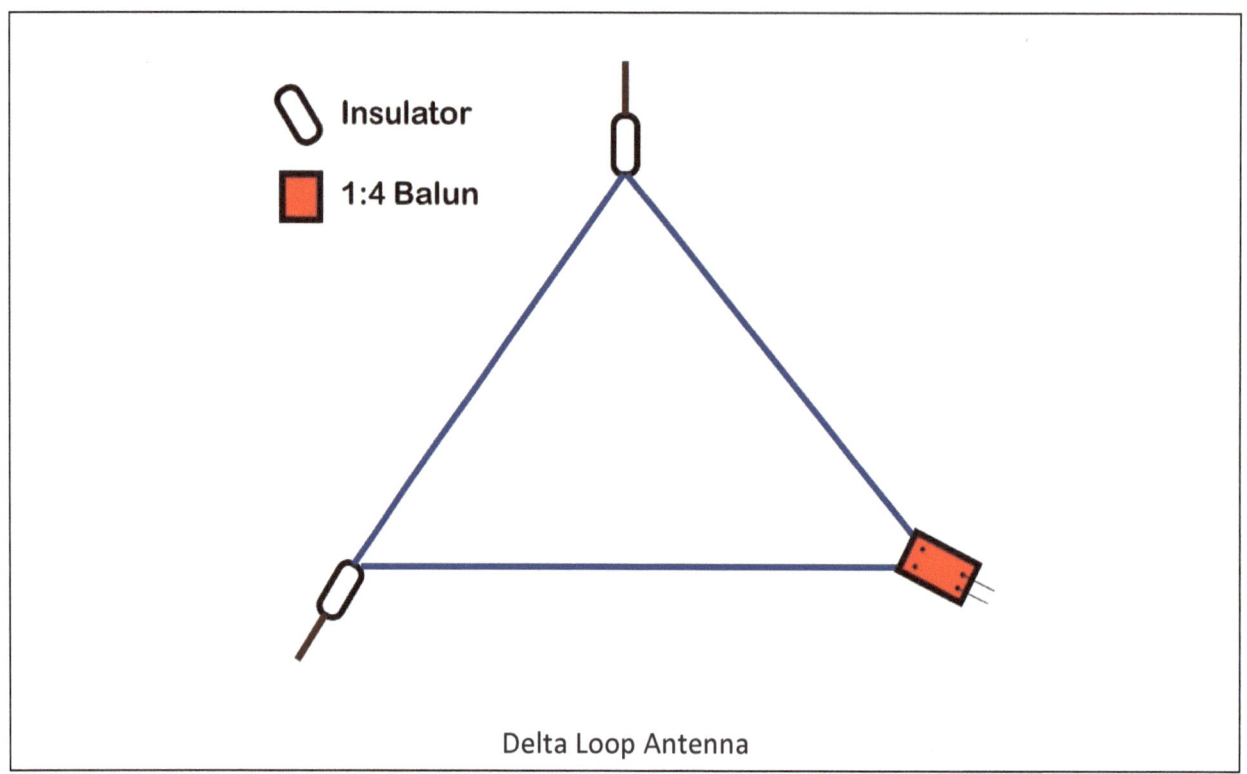

Delta Loop Antenna

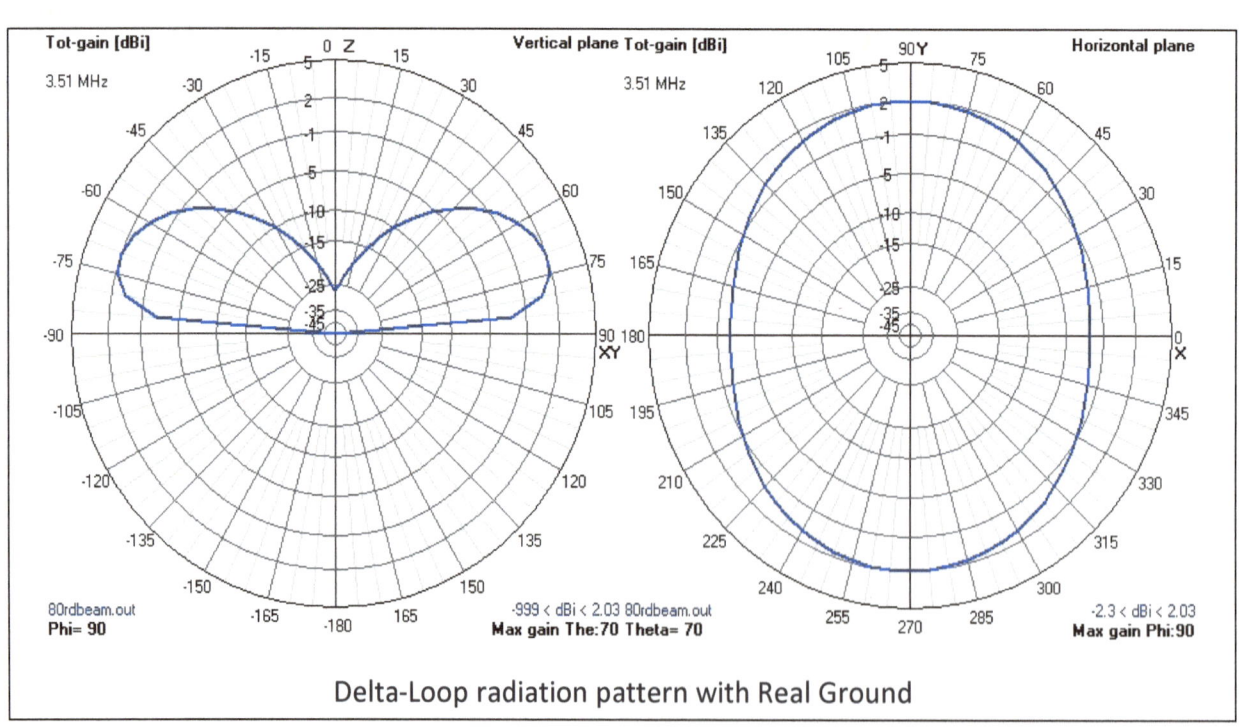

Delta-Loop radiation pattern with Real Ground

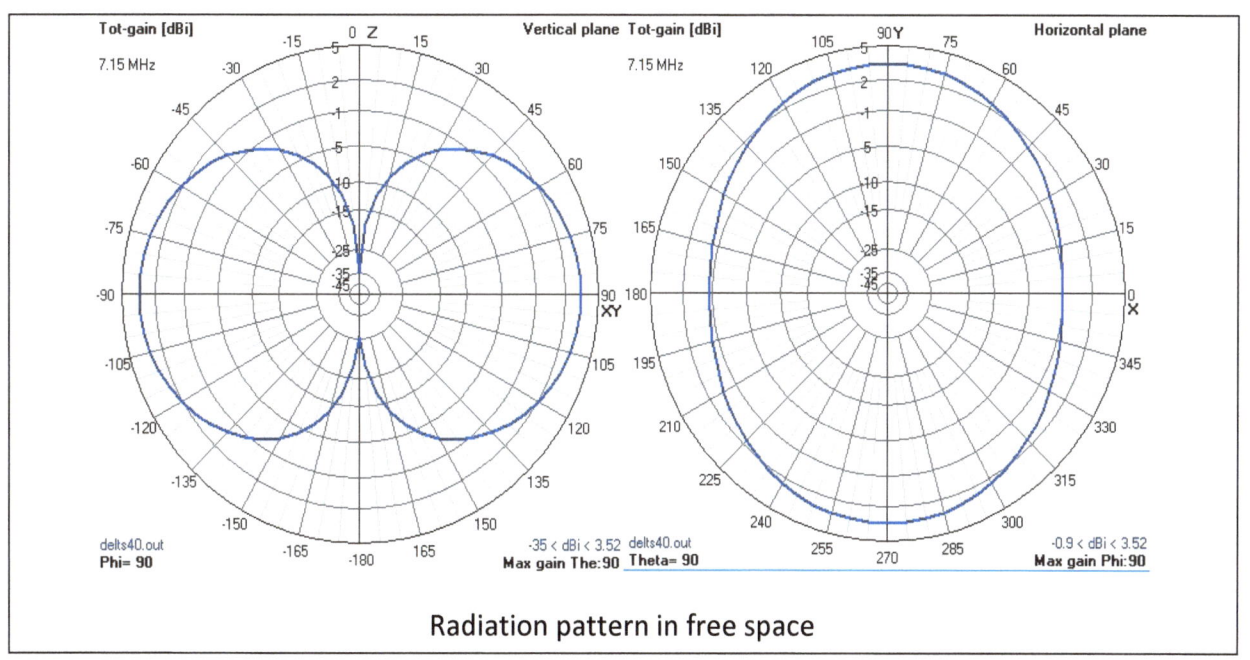

Radiation pattern in free space

Data Table

Band [m]	f [MHz]	Dipole length [m]	Arm length [m]	Inverted V length [m]	One arm length [m]	Delta Loop length [m]	One side [m]
160	1.9	75	37.5	71.3	35.65	157.2	52.4
80	3.5	40.8	20.4	38.7	19.35	78.17	26.06
60	5	28.5	14.25	27.1	13.55	61.26	20.42
40	7	20.4	10.20	19.4	9.7	40.79	13.60
30	10	14.3	7.15	13.5	6.75	28.68	9.56
20	14	10.2	5.10	9.7	4.85	20.41	6.80
17	18	7.9	3.95	7.5	3.75	17.01	5.67
15	21	6.8	3.4	6.5	3.25	13.65	4.55
12	24	5.94	2.97	5.64	2.82	11.66	3.88
11	27	5.28	2.64	5.02	2.51	11.35	3.78
10	28	5.09	2.545	4.84	2.42	10.24	3.41
6	50	2.85	1.425	2.71	1.355		
3	100	1.426	0.713	1.354	0.677		
2	144	0.990	0.495	0.940	0.47		
0.7	433	0.329	0.1645	0.312	0.156		
0.23	1290	0.110	0.055	0.104	0.052		
0.13	2450	0.058	0.029	0.055	0.0275		
0.06	5800	0.025	0.0125	0.023	0.0115		

Ground Plane (GP)

For a monopole antenna, the Earth – ground - reflects part of the waves, and it seems that they are coming from an image antenna – the other part of the dipole.

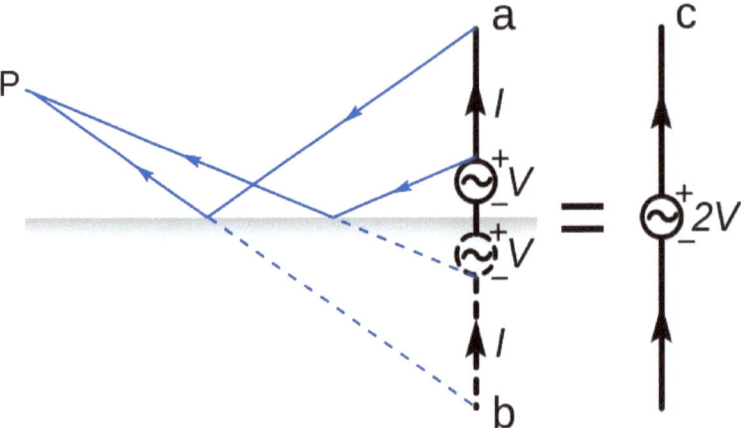

GP antennas do exactly that. Since we position antennas the highest we can, there must be a conductive element that acts as a reflective ground. For this purpose we can use an approximation of the ideal ground with 4 rods, each of the length of ¼ wavelength.

Impedance of the antenna will depend on the angle of the rods that imitate ground. Horizontal rods will give 35Ω, vertical "sleeve" will give as predicted (like vertical dipole) 75 Ω, and at the angle of approximately 40°, we will get the desired 50Ω.

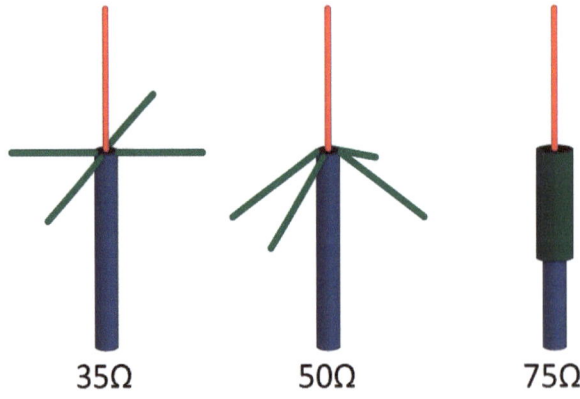

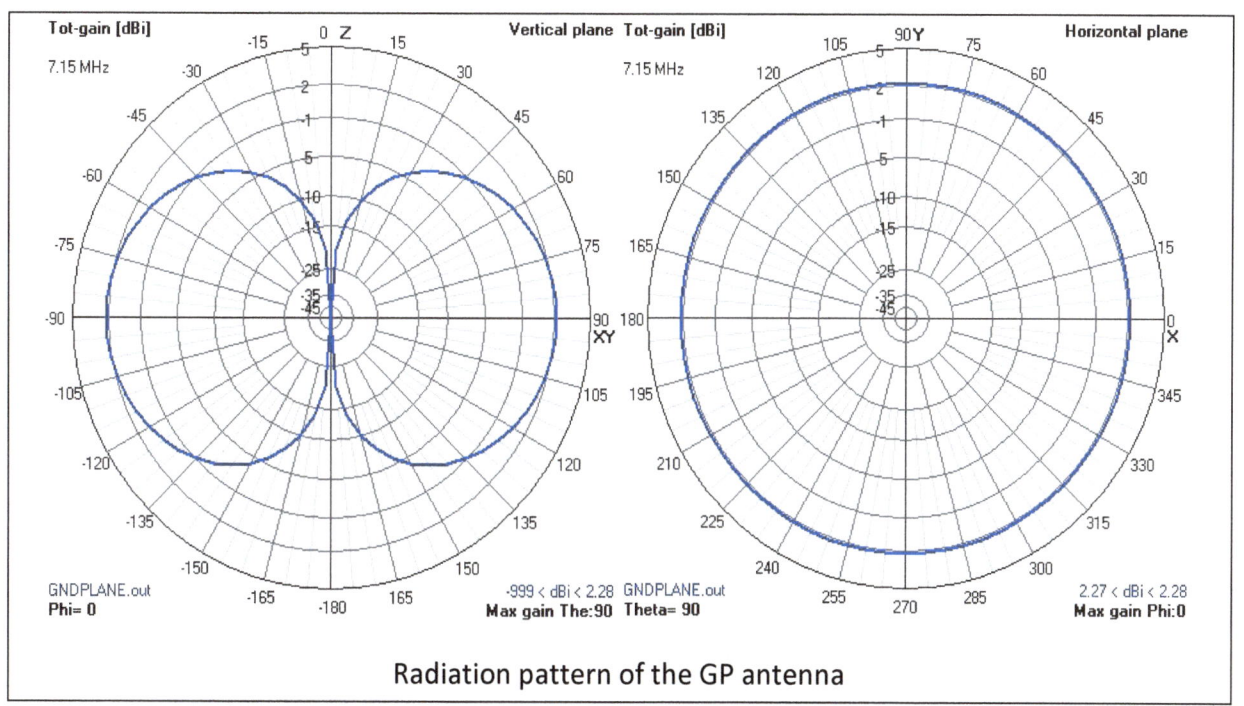

Radiation pattern of the GP antenna

Band	Frequency [MHz]	Ground radial [mm]	Vertical radial [mm]
6m	51	147	140
FM	100	75	71
2m	145	52	49
70cm	435	17	16

Construction
6m, FM and 2m

1. Make steel part like on the image below. The four ends are threaded with Ø5mm. The hole on the side is also with thread, diameter is optional.

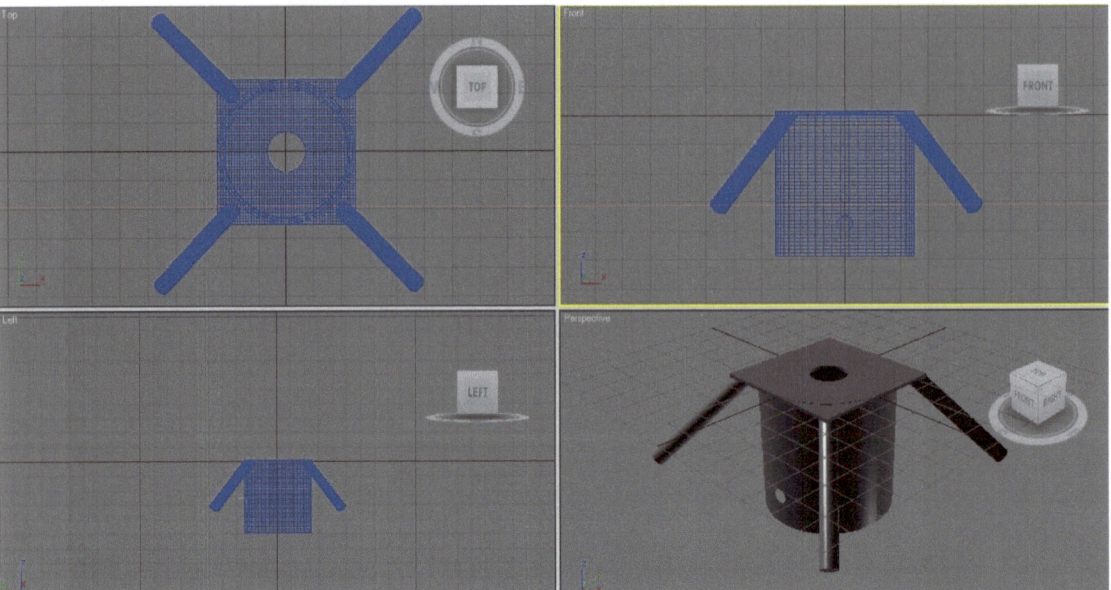

2. Cut the four aluminum Ø6mm pipes to the length (a bit longer than in the table above). Four longer pipes (ground) need to be threaded inside. To strengthen the part where the thread is cut four parts of Ø8mm aluminum pipe, with 1mm thick wall, a few centimeters long, and put them over the end of the Ø6mm pipe where the thread will be made. Create the thread in each pipe. Paint everything but the threaded parts.

3. Attach SO-239 or N connector through the large hole.

4. Cut the fifth pipe (from brass this time) a bit longer than at the table above. Cut one end into the pipe twice across its diameter, so that it looks like X. Then use pliers to wrap the pipe around the brass part of the connector (hot wire) and weld them together.

5. Now that the pipe stands up, we need a plastic part as on image – a nail for Styrofoam (easily found in hardware stores) does the trick. Screw it to the base part so that it is around the brass pipe. Use heat-shrinks to tighten everything up.

6. Use hot-glue gun to fill the hole at the top of the brass pipe.

7. Screw aluminum pipes to the steel base.

8. Cut all five pipes to the length given in the table above.

Photos of a finished and mounted GP antenna for 2m band

70cm

Note that this same design can be used for 2m, however the above method is better in every aspect.

1. Cut the five parts from the 3mm brass wire, few centimeters longer than in table above.
2. Make the metal part like on image, paint it and put the SO-239 or N connector (with four holes).
3. Bend brass wire parts that are for the ground, so that they can be held by screws.
4. Weld hot wire to the connector and isolate with hot-glue gun.
5. Assemble everything like on image.

Photo of a finished and mounted GP antenna for 70cm band

Vertical monopole with continuous loaded coil

This antenna type has the only advantage at being much smaller than ¼ wavelength monopole. Monopole at 160m would require 40m high antenna! Main disadvantage is its low efficiency; the larger the wavelength the higher are the losses. At 160m the losses are very high, but this is the only way to get antenna that can be mobile for that band.

The mast must be an insulator.

Height of the complete antenna is only 2190mm, for all the frequencies. Coil length is 2116mm, and there is the part of the mast at the bottom where wire from the coil goes straight down to the end of the mast (vertical green wire at the image below).

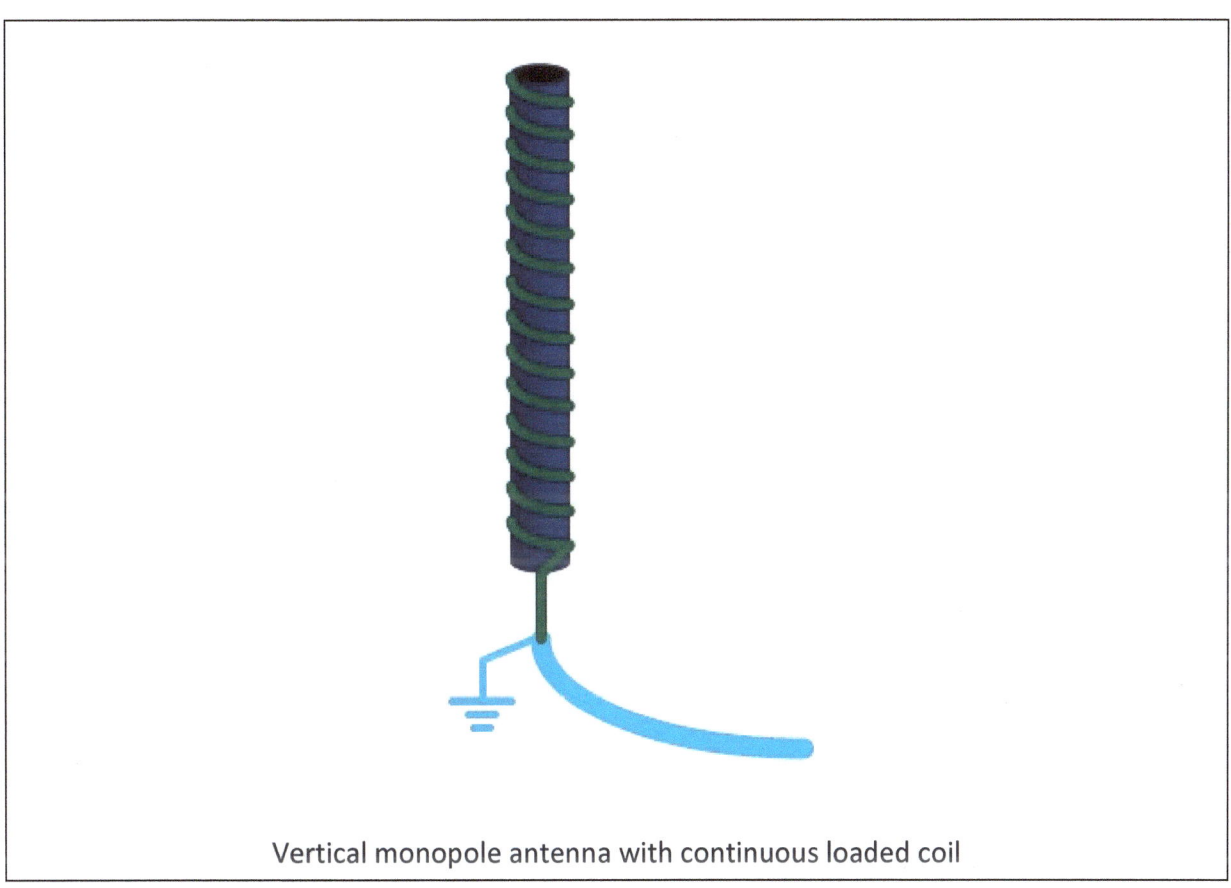

Vertical monopole antenna with continuous loaded coil

Band [m]	No. of turns	Coil diameter [mm]	Space between windings [mm]	Wire diameter [mm]	Loss [dB]
160	753	51	2.9	2.59	21.6
80	824	25	2.7	2.05	15
40	405	25	5.4	2.59	5.5
20	191	25	11.5	2.59	1

Yagi with Folded Dipole

This is the most common design for Yagi antennas. Folded dipole has advantages over the dipole in greater bandwidth and higher impedance, and it can be more easily adjusted with a BalUn to 50Ω required by the most devices[3].

For all the frequencies the principle is the same: there is one reflector that is behind dipole. Reflector has largest length of all the elements. After dipole come directors, each director further from the dipole is a bit shorter than the previous. Boom is the pipe that holds all the elements.

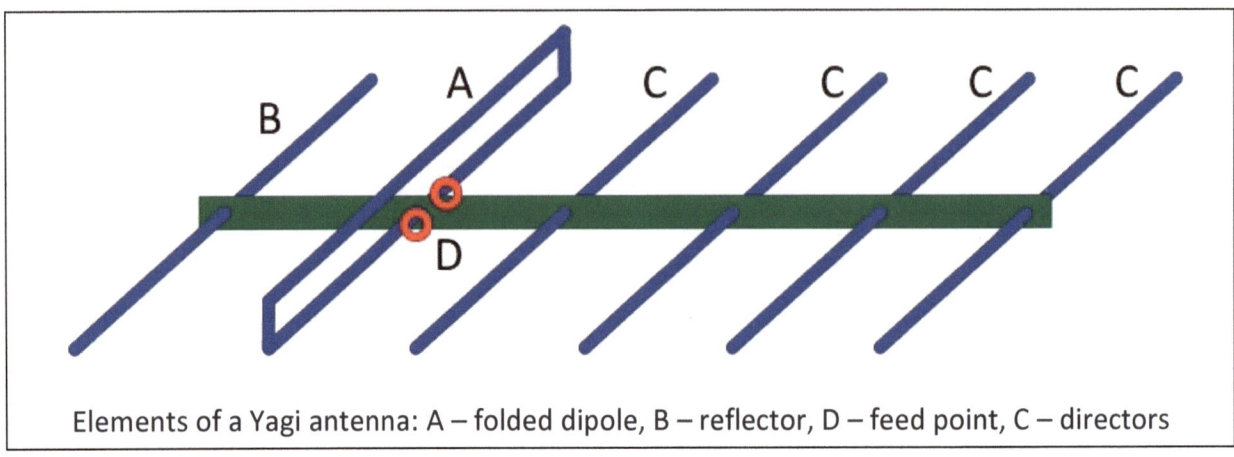

Elements of a Yagi antenna: A – folded dipole, B – reflector, D – feed point, C – directors

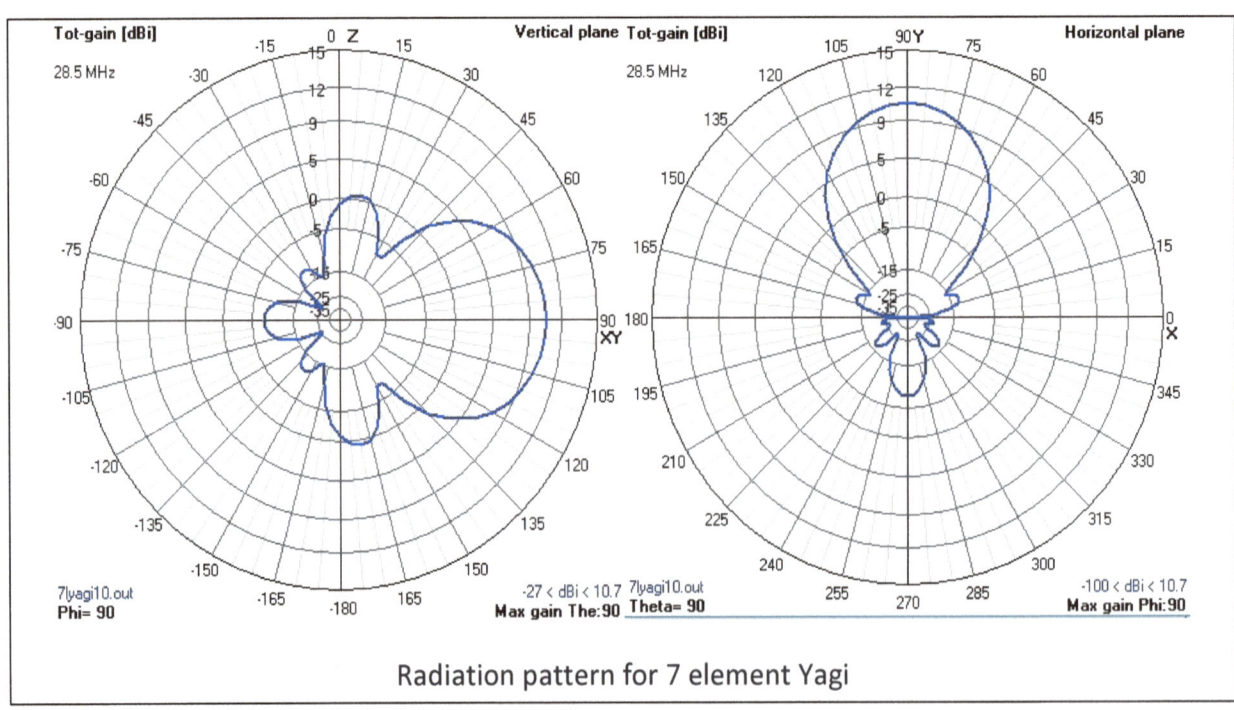

Radiation pattern for 7 element Yagi

---

[3] This is not always correct. In many cases 28Ω antennas are a better choice.

Construction

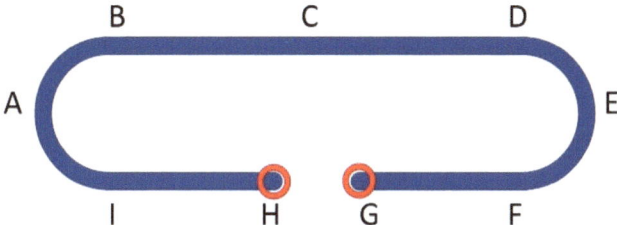

- "Elements" is the number of elements chosen for the band. Antenna can have less elements than the values provided, but it will result in less gain of the antenna. It can also have more directors, but for the practical reasons this book limits number of elements to the reasonable number for easy construction, depending on the band.
- "Diameter of dipole bend" is the outer diameter (in mm) of the aluminum pipe used – distance BI or FD.
- "Dipole gap at feed point" is the distance H-G.
- "Cross section of boom" is the outer diameter (or the length of one side of a rectangle if boom has square cross section).
- "Square or round boom" – the boom cross section type is chosen by the practicality for the band. Round boom is much stronger for the same width, but it is harder to drill accurately and harder to mount.
- "Director Shape" – same as with boom.
- "Director mounting" is very important parameter. If it is "bonded through metal boom" it means that directors are in direct contact with the boom. "Nonmetal" means that there is a director holder – a standoff made from special plastic[4], and the director is then insulated from the boom. The difference is in the length of the directors. If director is in direct contact with the boom, its length is increased by the width of the boom in comparison with the insulated director. The same goes for the Reflector.
- "Diameter of element" – all elements in this design have the same pipe outer diameter (in mm).
- "Dipole shape" – shape of the dipole can be different from the shape of the director. For the higher frequencies it may be more practical to use flat ribbon (a flat piece of metal) than the pipe or wire.
- "Width of dipole" – if the dipole is round, the width is the pipe's outer diameter, if the dipole is flat ribbon, the width is the second largest dimension of the metal piece. Thickness of the aluminum pipe or flat ribbon is of much less importance to be taken into

---

[4] That can withstand various weather conditions.

account, however it must not be too thin, because it needs to withstand harsh weather conditions.

- "Folded dipole length" – measure from tip to tip.
- "Total rod length" – the length of the pipe or flat ribbon when straightened up (H-G on the diagram).
- "Centre of rod" – one half of the "total rod length".
- "BalUn" – length of the delay line 4:1 BalUn (see more in the section "BalUn"). This is the length of the U shaped part of the 50Ω coaxial cable.

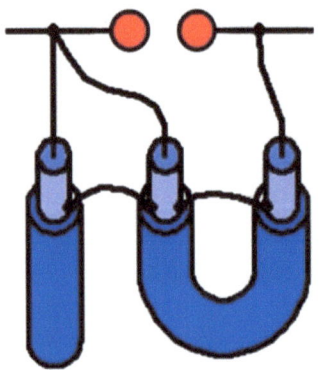

- "Length" – length of the element, tip to tip.
- "Spaced" – distance from the next element, center to center.
- "Boom position" – position where the element is placed on the boom.
- "Insert to" – mark the provided length to element, and place it on or through boom up to the marked point in order to center the element.
- "Gain" – if that number of elements is used, the maximum gain the antenna can have.
- "Boom" – boom length with added 30mm on each side after the last element and before the reflector.

Procedure

1. Cut the elements to the lengths as provided in table.
2. Cut the boom and mark the points where elements will be mounted.
3. Use plastic box as mount point for dipole, connector and to hold BalUn.
4. If the elements are mounted through boom, drill holes on the side of the boom with diameter that matches the element's diameter. If the elements are isolated from the boom, drill holes where the standoffs will be mounted.
5. Drill hole where the plastic box will mount to the boom.
6. On all the directors and reflector close the pipe ends with hot glue gun so that the water cannot enter the pipes.
7. Mark on all directors and reflector "insert to" points.

8. Insert all the directors and reflector up to the "insert to" point.
9. This step is only if the directors are bonded through metal boom. To ensure that the elements stay on their place, drill small holes into the boom (with inserted elements) and use screws for 1mm larger than the drill. The screws will create their path and hold tight. For each element use two screws.
10. Mark the points on the dipole. For pipe dipoles, use open flame to heat the pipe from point I to point B and from D to F. Do not overheat the pipe as it will deform the dipole. When heated, use the steel pipe that matches the inner diameter of the dipole bend and curve the pipe over it.
11. Flatten the ends of the dipole and drill small holes through the flattened part, closest to the edge as you can get.
12. Drill the holes into the plastic box where two sides of the dipole will enter and for the connector.
13. Mount the connector and dipole, insert BalUn, connect it to dipole and connector (weld it to connector, use screws to connect wire to aluminum dipole), and seal everything with hot-glue gun.
14. Mount the dipole and plastic box to the boom with a screw and close the box.
15. Mount antenna holder to the boom.

Data tables

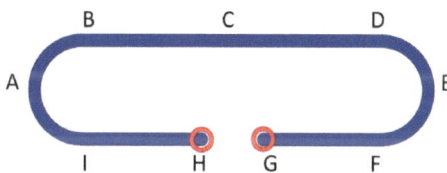

Band	f [MHz]	Elements	BI	HG	Cross section of boom	Boom shape	Director shape
FM	100	10	60	20	40	Square	Round
2	145	13	60	20	40	Square	Round
0.7	435	15	50	10	30	Square	Round
GSM	850	15	30	10	20	Square	Round
GSM	920	15	30	10	20	Square	Round
0.23	1296	29	30	10	20	Square	Round
GSM	1800	15	20	10	14	Square	Round
0.15	1900	15	20	10	14	Square	Round
0.13	2450	20	10	5	20	Round	Round

Band	f [MHz]	Director mounting	Diameter [mm]	Dipole shape	Width of dipole
FM	100	bonded through metal boom	8	Round	10
2m	145	bonded through metal boom	6	Round	6
70cm	435	bonded through metal boom	6	Round	6
GSM	850	bonded through metal boom	6	Flat ribbon	10
GSM	920	bonded through metal boom	6	Flat ribbon	10
23cm	1296	bonded through metal boom	6	Flat ribbon	10
GSM	1800	bonded through metal boom	4	Flat ribbon	7
GSM	1900	bonded through metal boom	4	Flat ribbon	7
Wi-Fi	2450	non metal	3	Flat ribbon	5

Band	f [MHz]	AE	Total length	HC	BC=CD	HI=GF	HA=GE	HB=GD	BalUn
FM	100	1447	2942	1471	693	683	731	778	989
2m	145	999	2047	1024	470	460	507	554	682
70cm	435	328	704	352	139	134	173	213	227
GSM	850	167	363	182	66	61	88	116	116
GSM	920	153	330	165	61	56	80	103	108
23cm	1296	107	239	119	39	34	57	81	76
GSM	1800	78	179	89	26	24	43	63	55
GSM	1900	77	173	86	26	21	41	60	52
Wi-Fi	2450	57	120	60	23	21	29	36	40

100MHz

100MHz	Length	Spaced	Boom position	Insert to	Gain [dBi]	Boom
Reflector	1482	0	30	721	0	60
Dipole	1447	600	630	703.5	4.3	660
1	1363.1	224.8	854.4	661.5	6.9	884.4
2	1350.4	539.6	1394.1	655	8.6	1424.1
3	1338.7	644.6	2038.6	649.5	9.9	2068.6
4	1327.8	749.5	2788.1	644	11	2818.1
5	1317.8	839.4	3627.5	639	11.9	3657.5
6	1308.4	899.4	4526.9	634	12.7	4556.9
7	1299.8	944.3	5471.2	630	13.3	5501.2
8	1291.7	989.3	6460.5	626	13.9	6490.5

145MHz

145MHz	Length	Spaced	Boom position	Insert to	Gain [dBi]	Boom
Reflector	1031	0	30	495.5	0	60
Dipole	999	414	444	497.5	4.3	474
1	946.2	155.1	598.6	453	6.9	628.6
2	937.4	372.2	970.7	448.5	8.6	1000.7
3	929.2	444.5	1415.2	444.5	9.9	1445.2
4	921.6	516.9	1932.1	441	11	1962.1
5	914.5	578.9	2511	437.5	11.9	2541
6	908	620.3	3131.3	434	12.7	3161.3
7	901.9	651.3	3782.6	431	13.3	3812.6
8	896.3	682.3	4464.9	428	13.9	4494.9
9	891.1	713.3	5178.2	425.5	14.4	5208.2
10	886.3	744.3	5922.5	423	14.9	5952.5
11	881.8	775.3	6697.8	421	15.3	6727.8

435MHz

435MHz	Length	Spaced	Boom position	Insert to	Gain [dBi]	Boom
Reflector	358	0	30	164	0	60
Dipole	149	138	168	149	4.3	198
1	318.4	51.7	219.5	144	6.9	249.5
2	314.8	124.1	343.6	142.5	8.6	373.6
3	311.4	148.2	491.7	140.5	9.9	521.7
4	308.3	172.3	664	139	11	694
5	305.5	193	857	137.5	11.9	887
6	302.8	206.8	1063.8	136.5	12.7	1093.8

7	300.4	217.1	1280.9	135	13.3	1310.9
8	298.1	227.4	1508.3	134	13.9	1538.3
9	296	237.8	1746.1	133	14.4	1776.1
10	294	248.1	1994.2	132	14.9	2024.2
11	292.2	258.4	2252.6	131	15.3	2282.6
12	290.5	265.3	2517.9	130.5	15.7	2547.9
13	289	268.8	2786.7	129.5	16	2816.7

850MHz

850MHz	Length	Spaced	Boom position	Insert to	Gain [dBi]	Boom
Reflector	185.9	0	30	83	0	60
Dipole	165.7	71	101	71	4.3	131
1	161.3	26.5	127	70.5	6.9	157
2	159.2	63.5	190.5	69.5	8.6	220.5
3	157.3	75.8	266.3	68.5	9.9	296.3
4	155.5	88.2	354.5	68	11	384.5
5	153.9	98.8	453.2	67	11.9	483.2
6	152.4	105.8	559	66	12.7	589
7	150.9	111.1	670.1	65.5	13.3	700.1
8	149.6	116.4	786.5	65	13.9	816.5
9	148.4	121.7	908.2	64	14.4	938.2
10	147.3	127	1035.2	63.5	14.9	1065.2
11	146.2	132.3	1167.4	63	15.3	1197.4
12	145.2	135.8	1303.2	62.5	15.7	1333.2
13	144.3	137.6	1440.8	62	16	1470.8

920MHz

920MHz	Length	Spaced	Boom position	Insert to	Gain [dBi]	Boom
Reflector	172.8	0	30	76.5	0	60
Dipole	152.7	65	95	66.5	4.3	125
1	149.6	24.4	119.6	65	6.9	149.6
2	147.7	58.7	178.3	64	8.6	208.3
3	145.9	70.1	248.3	63	9.9	278.3
4	144.2	81.5	329.8	62	11	359.8
5	142.7	91.2	421	61.5	11.9	451
6	141.2	97.8	518.8	60.5	12.7	548.8
7	139.9	102.6	621.4	60	13.3	651.4
8	138.6	107.5	729	59.5	13.9	759
9	137.5	112.4	841.4	59	14.4	871.4

10	136.4	117.3	958.7	58	14.9	988.7
11	135.5	122.2	1080.9	57.5	15.3	1110.9
12	134.6	125.5	1206.4	57.5	15.7	1236.4
13	133.7	127.1	1333.4	57	16	1363.4
14	132.9	128.7	1462.2	56.5	16.3	1492.2
15	132.2	130.3	1592.5	56	16.6	1622.5

1296MHz

1296MHz	Length	Spaced	Boom position	Insert to	Gain [dBi]	Boom
Reflector	126.7	0	30	53.5	0	60
Dipole	107.2	46	76	43.5	4.3	106
1	108.8	17.3	93.6	44.5	6.9	123.6
2	107.3	41.6	135.3	43.5	8.6	165.3
3	105.9	49.7	185	43	9.9	215
4	104.7	57.8	242.8	42.5	11	272.8
5	103.5	64.8	307.6	42	11.9	337.6
6	102.4	69.4	377	41	12.7	407
7	101.4	72.9	449.8	40.5	13.3	479.8
8	100.5	76.3	526.2	40	13.9	556.2
9	99.6	79.8	606	40	14.4	636
10	98.8	83.3	689.3	39.5	14.9	719.3
11	98.1	86.7	776	39	15.3	806
12	97.4	89.1	865.1	38.5	15.7	895.1
13	96.7	90.2	955.3	38.5	16	985.3
14	96.2	91.4	1046.7	38	16.3	1076.7
15	95.6	92.5	1139.2	38	16.6	1169.2
16	95.1	92.5	1231.7	37.5	16.9	1261.7
17	94.6	92.5	1324.2	37.5	17.1	1354.2
18	94.2	92.5	1416.8	37	17.4	1446.8
19	93.8	92.5	1509.3	37	17.6	1539.3
20	93.4	92.5	1601.8	36.5	17.8	1631.8
21	93.1	92.5	1694.4	36.5	18	1724.4
22	92.7	92.5	1786.9	36.5	18.2	1816.9
23	92.4	92.5	1879.4	36	18.3	1909.4
24	92.2	92.5	1971.9	36	18.5	2001.9
25	91.9	92.5	2064.5	36	18.7	2094.5
26	91.7	92.5	2157	36	18.8	2187
27	91.4	92.5	2249.5	35.5	19	2279.5

1800MHz

1800MHz	Length	Spaced	Boom position	Insert to	Gain [dBi]
Reflector	91.8	0	30	38.5	0
Dipole	77.7	33	63	31.5	4.3
1	79.1	12.5	75.8	32	6.9
2	78.1	30	105.8	31.5	8.6
3	77.1	35.8	141.6	31	9.9
4	76.2	41.6	183.2	30.5	11
5	75.4	46.6	229.9	30	11.9
6	74.6	50	279.8	30	12.7
7	73.9	52.5	332.3	29.5	13.3
8	73.2	55	387.3	29	13.9
9	72.6	57.5	444.7	29	14.4
10	72.1	60	504.7	28.5	14.9
11	71.5	62.5	567.1	28.5	15.3
12	71	64.1	631.2	28	15.7
13	70.6	65	696.2	28	16

1900MHz

1900MHz	Length	Spaced	Boom position	Insert to	Gain [dBi]
Reflector	91	0	30	38.5	0
Dipole	77.1	33	63	31.5	4.3
1	78.3	12.5	75.8	32	6.9
2	77.3	30	105.8	31.5	8.6
3	76.3	35.8	141.6	31	9.9
4	75.4	41.6	183.2	30.5	11
5	74.6	46.6	229.9	30.5	11.9
6	73.8	50	279.8	30	12.7
7	73.1	52.5	332.3	29.5	13.3
8	72.4	55	387.3	29	13.9
9	71.8	57.5	444.7	29	14.4
10	71.3	60	504.7	28.5	14.9
11	70.7	62.5	567.1	28.5	15.3
12	70.3	64.1	631.2	28	15.7
13	69.8	65	696.2	28	16

2450MHz	Length	Spaced	Boom position	Insert to	Gain [dBi]
Reflector	59	0	30	19.5	0
Dipole	57	24	54	17.5	4.3
1	49.4	9.2	63.7	14.5	6.9
2	48.6	22	85.7	14.5	8.6
3	47.9	26.3	112	14	9.9
4	47.3	30.6	142.6	13.5	11
5	46.6	34.3	176.8	13.5	11.9
6	46.1	36.7	213.5	13	12.7
7	45.6	38.5	252.1	13	13.3
8	45.1	40.4	292.5	12.5	13.9
9	44.6	42.2	334.7	12.5	14.4
10	44.2	44.1	378.7	12	14.9
11	43.8	45.9	424.6	12	15.3
12	43.4	47.1	471.7	11.5	15.7
13	43.1	47.7	519.5	11.5	16
14	42.8	48.3	567.8	11.5	16.3
15	42.5	48.9	616.7	11.5	16.6
16	42.2	48.9	665.7	11	16.9
17	42	48.9	714.6	11	17.1
18	41.8	48.9	763.6	11	17.4

2-element Yagi

There is no antenna with better gain/expense ratio than the 2-element Yagi antenna. The Dipole alone gives us 2.15dBi. As we can see from the radiation pattern, the maximum gain (in the direction of the x-axes) is 6.5dBi, a gain in over 4dBi by just adding a reflector or a director! By adding one additional director we will have only approximately 1dBi more.

For the lower frequencies, where the dimension of the antenna becomes quite significant, as well as its price, it is obvious that adding that additional element does not pay off.

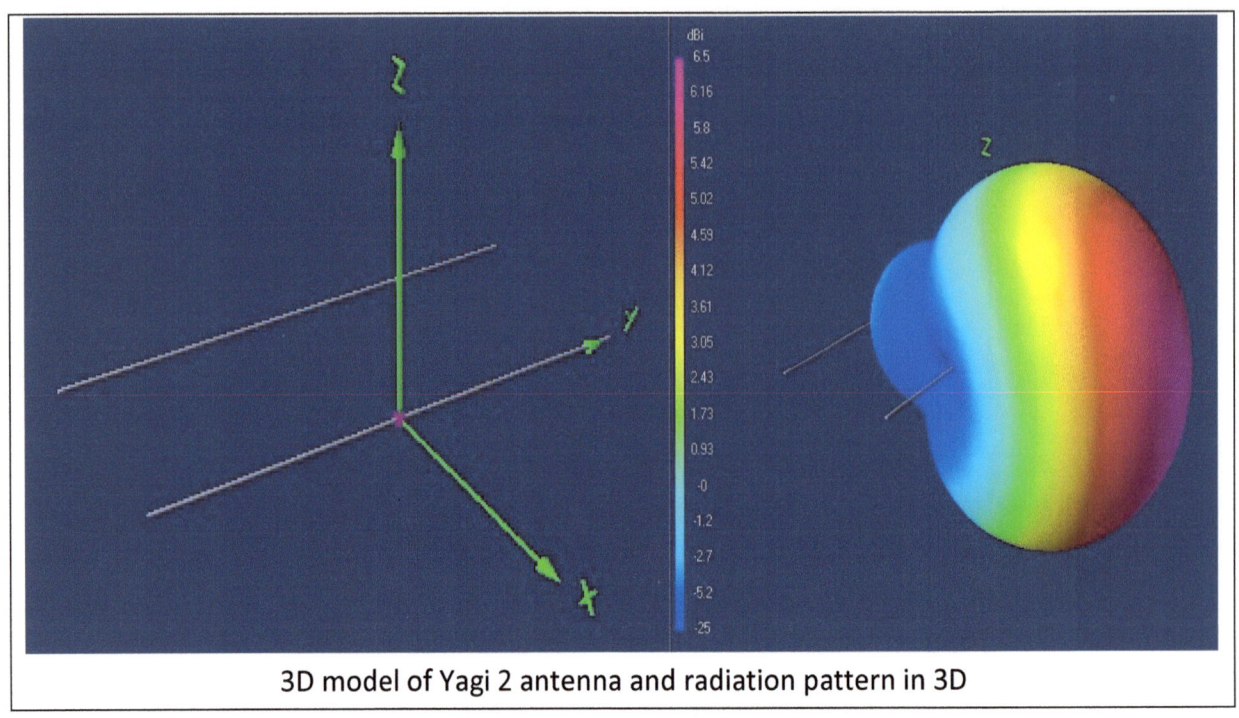

3D model of Yagi 2 antenna and radiation pattern in 3D

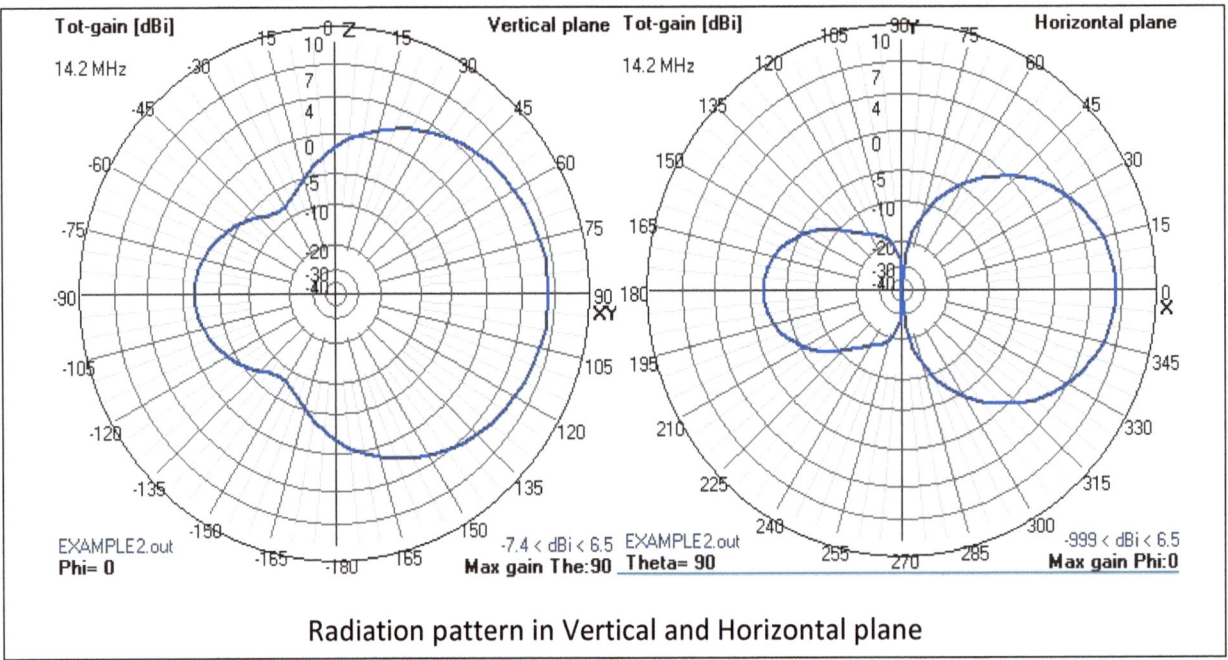

Radiation pattern in Vertical and Horizontal plane

There are two variants of the Yagi 2: with Reflector and without Directors and with one Director but without Reflector. If the value in the last column is larger than for the Radiator-Dipole, then it is the measure for the Reflector, and vice versa – Reflectors are always longer than the Dipole and the Directors are always shorter.

Band [m]	Impedance [Ω]	Boom [mm]	Radiator [mm]	Reflector or Director [mm]
10	50	1760	4930	5360
12	28	1180	6050	5600
12	12.5	670	6090	5780
15	28	1610	6870	7220
17	28	1700	8320	7800
20	28	2300	10260	11080
30	28	3000	14920	14040

There is a need for an appropriate BalUn. See more in the BalUn section.

HB9CV

This antenna is made from two elements but the gain is about the same as the 3-element Yagi. At lower frequencies, where the elements are quite large, one can see that this antenna has the huge advantage in price, weight and wind resistance. The gain is approximately 6.5dBi. The only small drawback is the need for a capacitor for tuning the SWR.

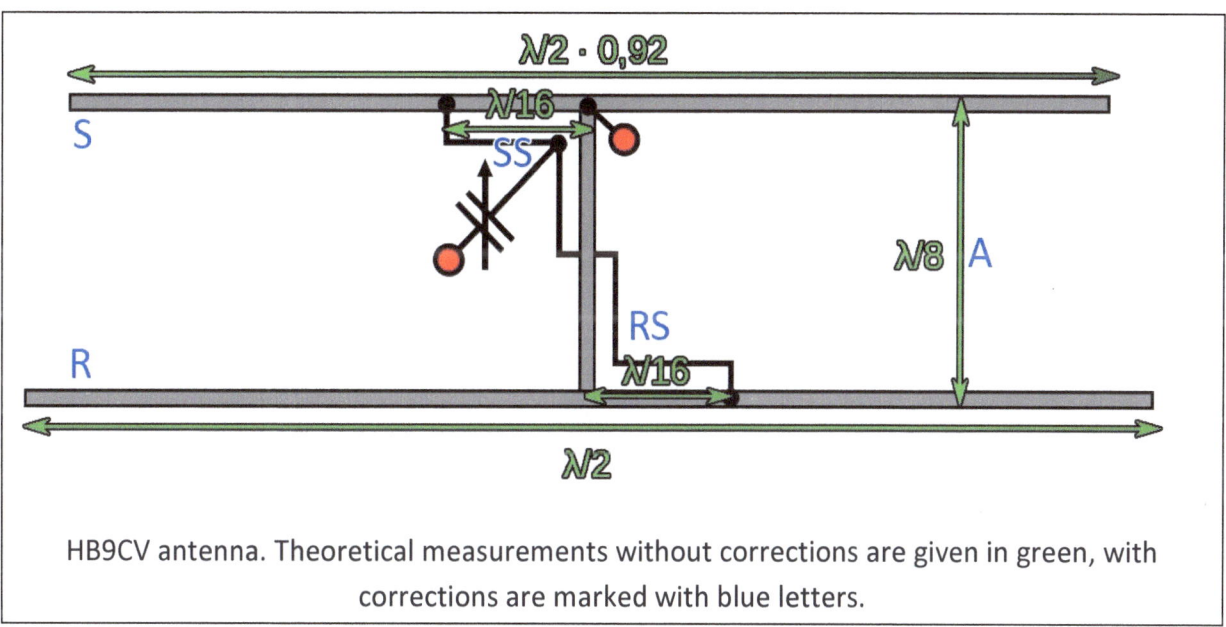

HB9CV antenna. Theoretical measurements without corrections are given in green, with corrections are marked with blue letters.

Put the capacitor on a hot-wire lead.

- R – Reflector length,
- S – Length of the Radiating element,
- A – distance between R and S, center to center,
- RS – distance from the line of the symmetry of the antenna and the point where the phasing line is attached to the R,

- SS – distance from the line of the symmetry of the antenna (line that passes through the center of the boom) and the point where the phasing line is attached to the S,
- C – Approximate capacity for which the best SWR can be achieved. It is best to use variable capacitor to find the exact capacity, and then replace it with fixed capacitor.

Band [m]	R [mm]	S [mm]	A [mm]	RS [mm]	SS [mm]	C [pF]
10	5300	4900	1330	800	760	56
6	3000	2770	750	450	430	30
2	1020	945	260	190	180	12

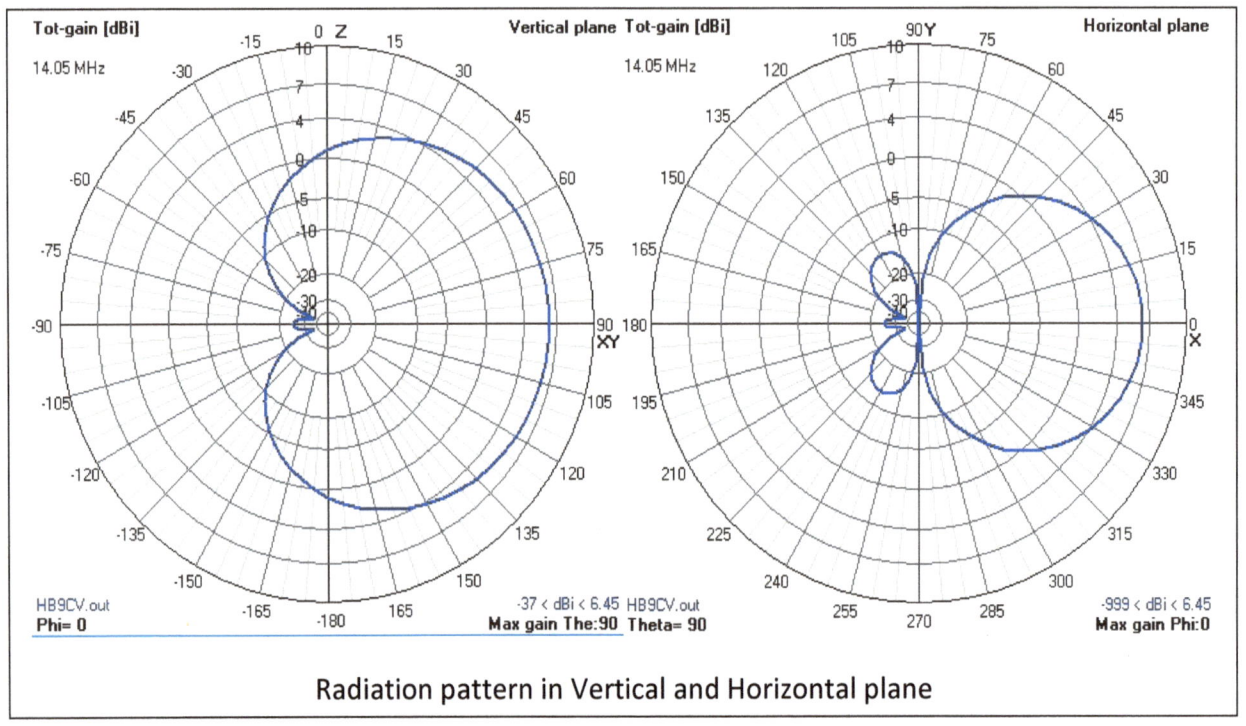

Radiation pattern in Vertical and Horizontal plane

Biquad

Biquad antenna is also known as Double Quad Beam or some call it Bowtie Beam antenna[5]. It is widely used because of its properties:

- Compact and thus can be portable,
- Easy to make,
- Cheap,
- Only one radiating element and reflector,
- 11dBi gain,
- SWR as low as 1.05 if made ideally,

---

[5] Bowtie Beam is also the name for the subtype of the Biconical antenna.

- 50Ω impedance,
- Equal radiation pattern in horizontal and vertical planes,
- Good bandwidth, better than Yagi.

Dimensions of the reflector are not crucial. By Trevor Marshal, it should be 0.9×0.9 of the wavelength, but it was discovered that 1.2×1.6 wavelengths gives the largest gain. In the table we will list both values.[6] Distance of radiating element from reflector is from the center of the wire.

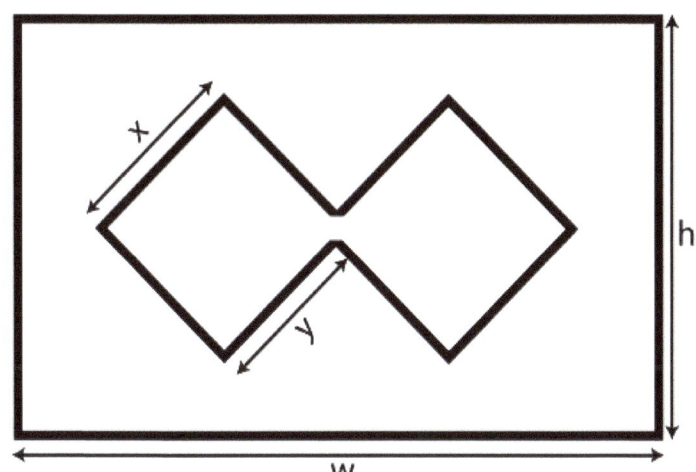

- w – Reflector width,
- h – Reflector height,
- x – Outer Side Length,
- y – Inner Side Length.
- All measurements are in *mm*.

Band	Frequency [MHz]	λ	w & h (0.9λ)	w (1.6λ)	h (1.2λ)	Reflector type
70cm	435	689	620	1100	827	Wireframe
23cm	1296	231	240	370	277	Wireframe
2.4GHz	2430	123	110	150	200	Aluminum plate
5GHz	5300	57	51	68	93	Aluminum plate

Band	Wire Length	x	y	Distance from Reflector	Wire diameter
70cm	1340	170	165	87	5
23cm	464	58	54	29	3
2.4GHz	241	31	29	15	2
5GHz	110	15	12.5	7	2

---

[6] The larger reflector will give us less than 1dBi gain over the smaller one.

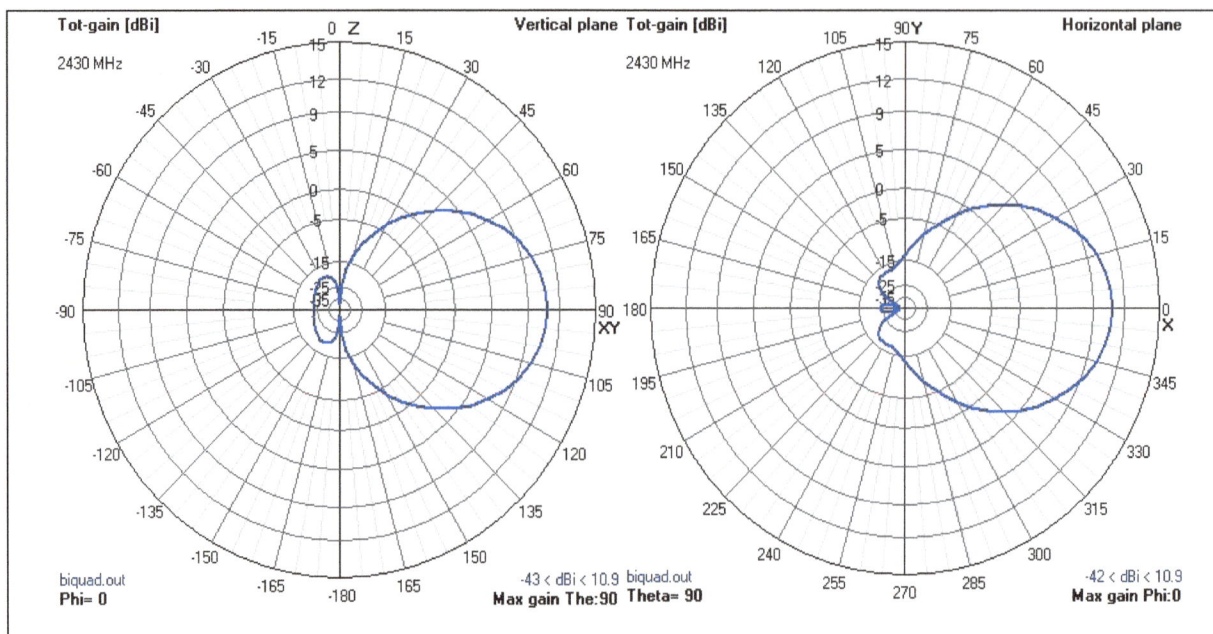

Radiation pattern in Vertical and Horizontal plane. Image below: in 3D

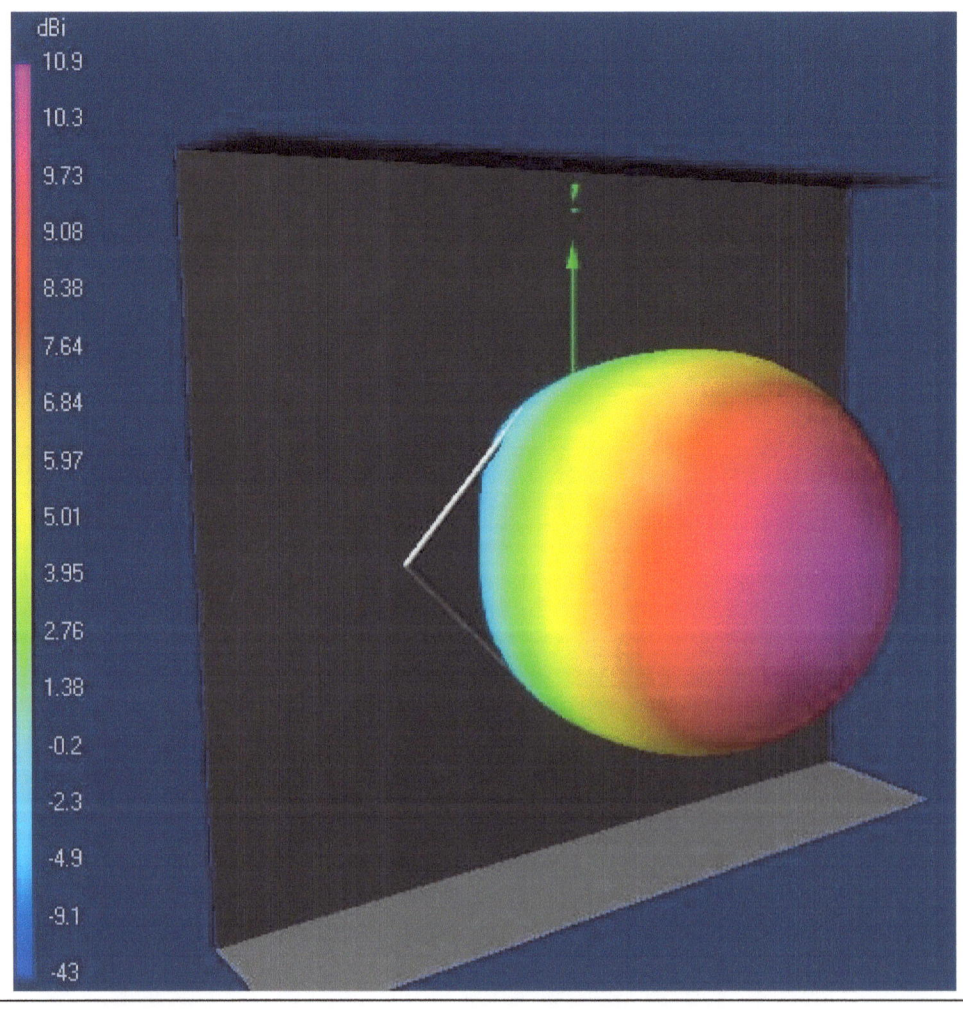

Orientation and polarization are not intuitive as with Yagi antenna. See the image for proper orientation.

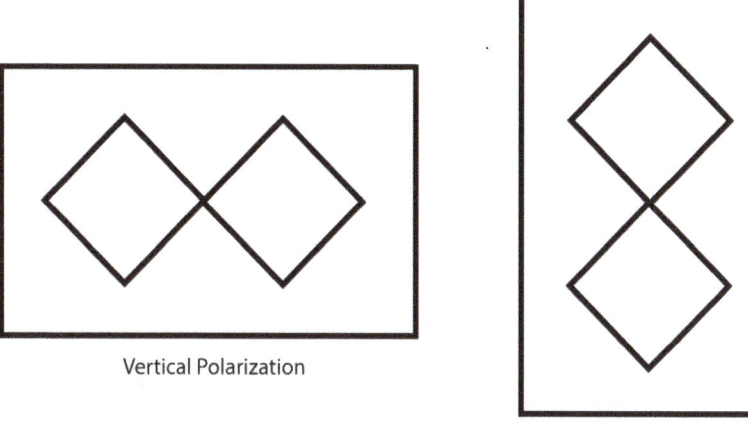

Vertical Polarization

Horizontal Polarization

3D Corner

3D corner antenna was designed in the 1970s, and is very often used, even for military radars. The design given here is a modification by *Dragoslav Dobricic*. The modifications from the original design are smaller antenna and lowering the impedance with one director[7] which has as a consequence improved gain.

The antenna is practical for many bands, with slight modifications. For higher frequencies, 3D corner is one of the best feeds for offset parabolic satellite dishes.

Its mechanism of work can be split into two parts: reflector and monopole with director.

Reflector is known as Corner Reflector – three perpendicular flat conducting surfaces that reflect waves. For the maximum gain, antenna should be tilted by 45°.

---

[7] The same principle is used with Yagi's.

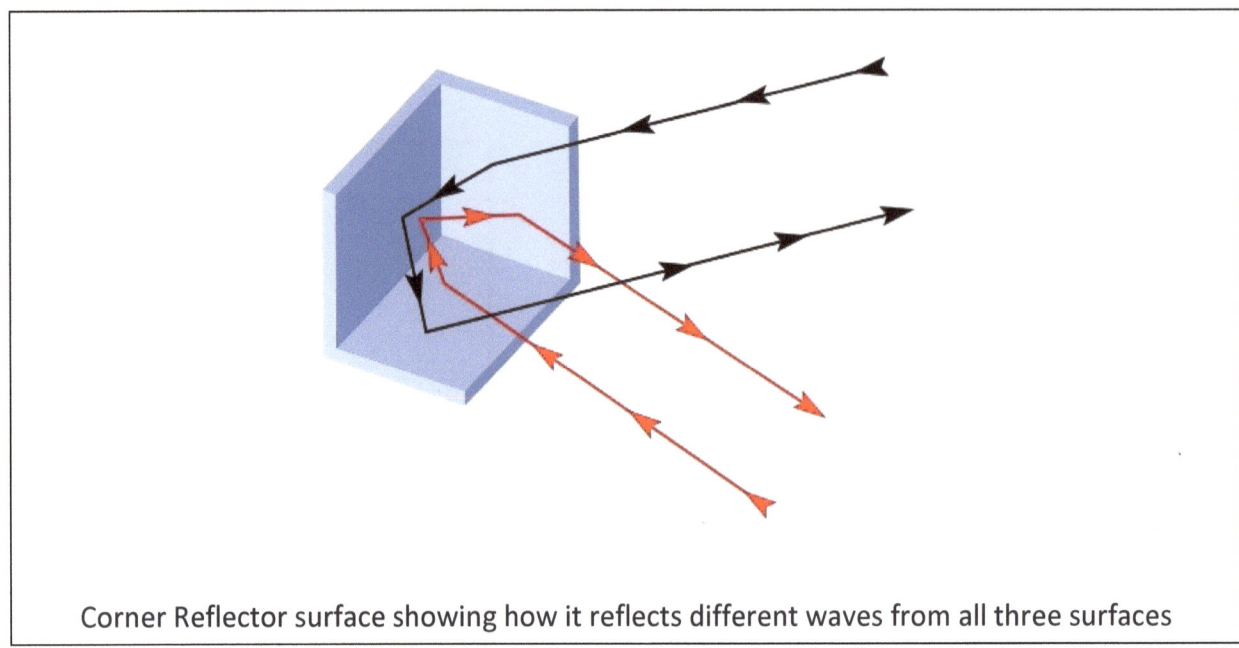

Corner Reflector surface showing how it reflects different waves from all three surfaces

Yagi-like part consists from a monopole and a director.

Reflector can be of any size, keeping in mind that its size directly affects gain. Its optimum size is at 2.8 wavelengths for one side, for which the gain is 18dBi. Larger reflectors do not give significant increase in gain. Very good 16dBi can be obtained with 1.8 wavelengths.

Reflector side [λ]	1.2	1.4	1.8	2.3	2.8	3.8
Gain [dBi]	12	14	16	17	18	18

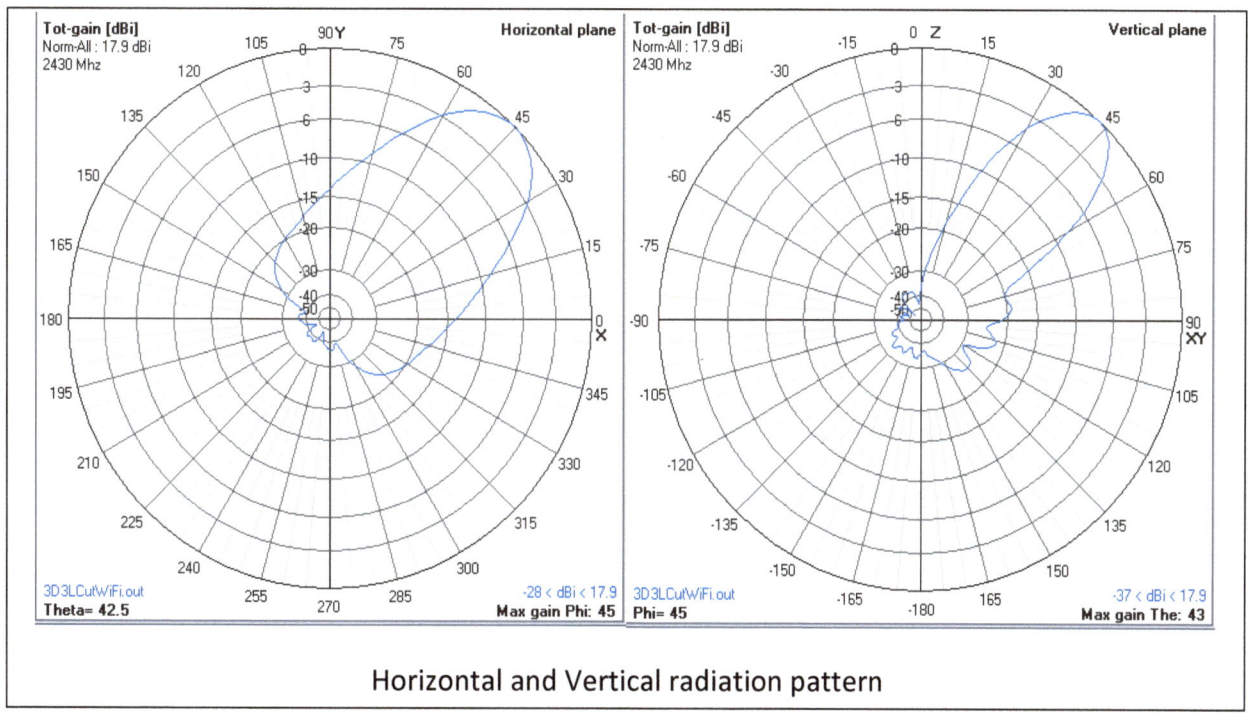

Horizontal and Vertical radiation pattern

Reflector should be made from 1mm thick aluminum plate for higher frequencies, or with net wire with density 0.1 wavelengths for 23cm band or lower frequencies. In some cases, because of the high wind resistance of the antenna, a metal frame is required.

Monopole should be placed on 0.6 wavelengths from both reflector surfaces, and should be 0.75 wavelengths long.

With monopole only and without the director, antenna has 72Ω impedance, and could be feed with cheap RG-6 cable. The only problem lies that all the devices are adjusted to work with 50Ω.

By adding a passive element – director – impedance would be lowered to 50Ω. Director should be placed 0.7 wavelengths from the reflector's surfaces. It should be 0.65 wavelengths long, measuring from the reflector surface.

Director and monopole should be connected by PVC pipe, which will ensure their stability.

To sum up dimensions for all the frequencies:

- For 18dBi, reflector should have 3λ sides,
- Monopole should be long 0.754λ, measured from reflector surface,
- Director should be 0.65λ, measured from reflector surface,
- Director and monopole thickness: 0.032λ,
- Director position: 0.7λ from reflector surfaces,
- Monopole position: 0.6λ from reflector surfaces.

Legend (all dimensions are in *mm*):

- m – monopole length,
- d – director length,
- w – net wire density,
- r – minimum reflector thickness,
- h – director and monopole thickness,
- x – monopole distance from reflector surfaces,
- y – director distance from reflector surfaces,
- R – length of a side of the reflector.

Band	Frequency [MHz]	Wavelength	Reflector			
			12dBi	14dBi	16dBi	18dBi
2m	145	2070	2484	2898	3726	6210
70cm	435	690	828	966	1242	2070
23cm	1296	230	276	322	414	690
2.4GHz	2430	123	147	172	221	369
5GHz	5600	54	65	76	97	162

It seems that antenna is not usable at lower frequencies, for example for 2m band, but that might not be true if we can make use of some existing construction. However, the reflector that gives over 12dBi is not likely to be used.

Band	m	d	w	r	h	y	x
2m	1561	1346	207	/	40	1449	1242
70cm	520	449	69	/	20	483	414
23cm	173	150	23	1.2	7	161	138
2.4GHz	93	80	/	1	4	86	74
5GHz	41	35	/	0.5	2	38	32

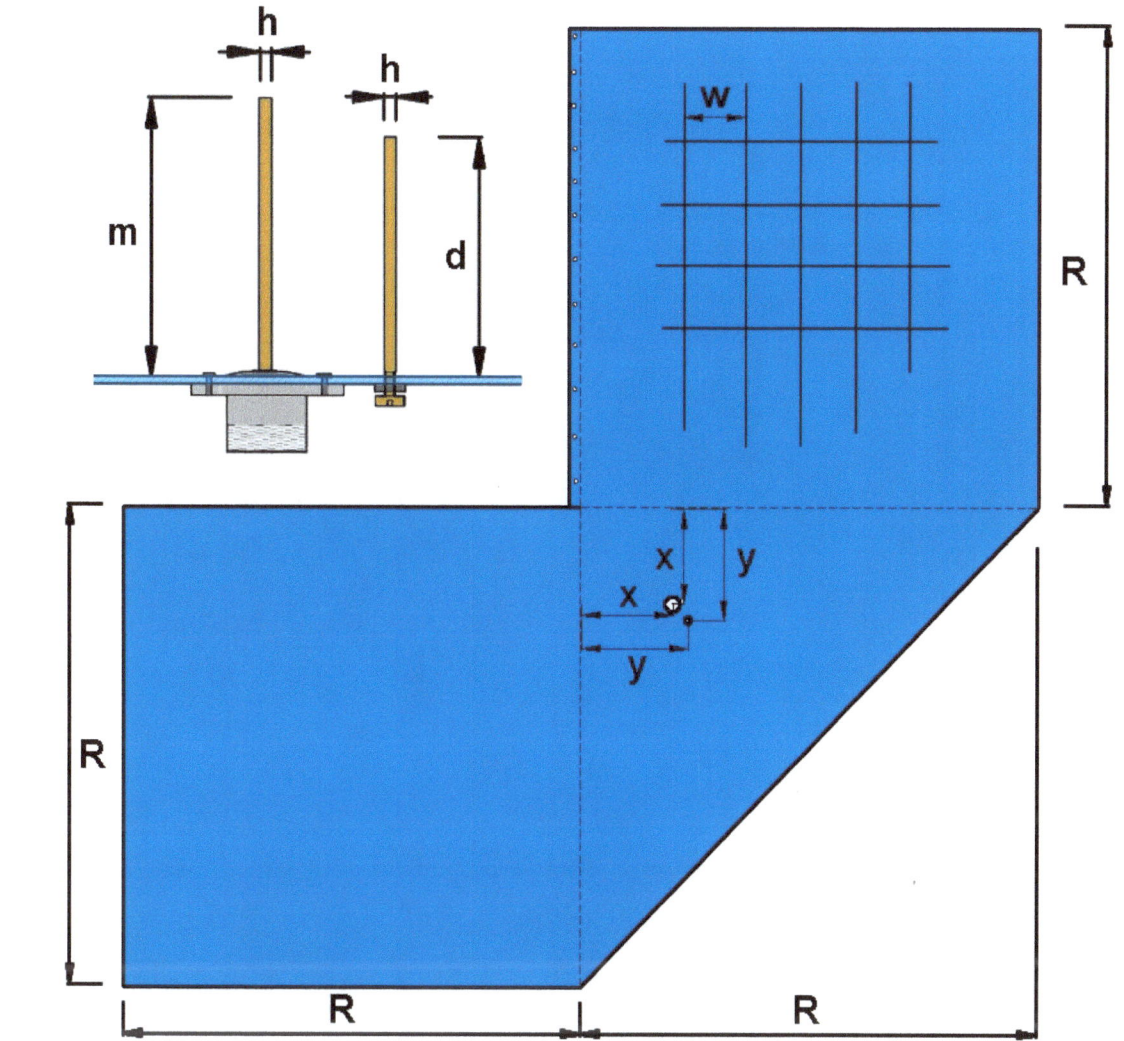

Photos of a 3D corner antenna for 2.4GHz and a 16dBi reflector

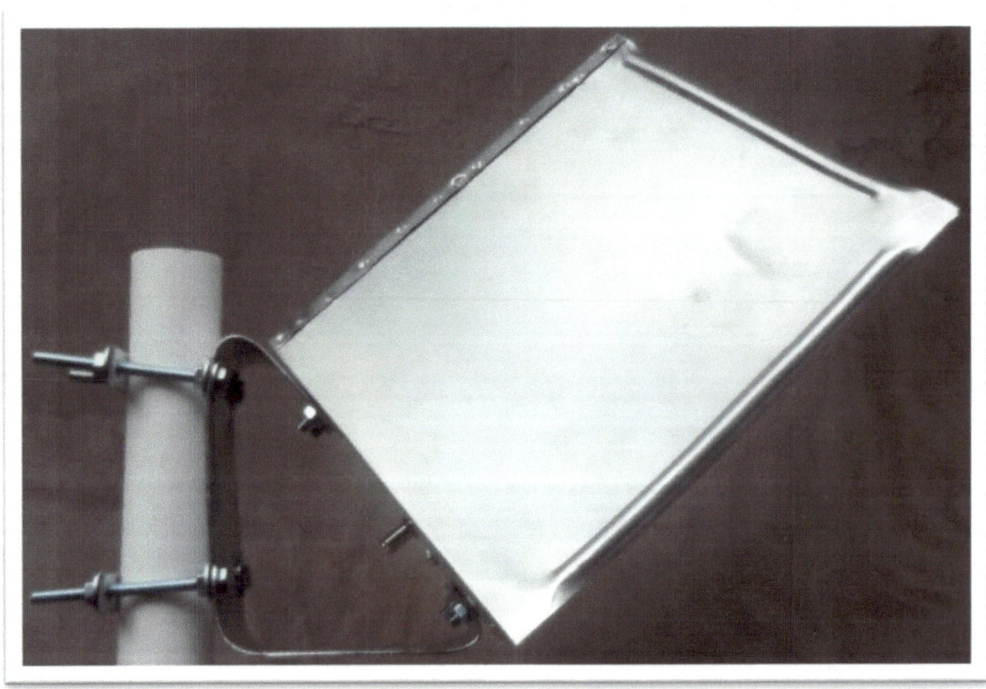

Radio broadcast

In this section we will describe antennas that are made for receiving TV signal, for FM broadcast, for FM receiving or for AM receiving. AM broadcast or TV broadcast antennas are not described in this book for the obvious reasons.

AM

AM band lies between 535 kHz and 1705 kHz. Although it is a dying band, there are some radio stations that can be found on it. Though propagation depends largely on the time of the day (at night the propagation is the best), some radio stations use so much power that they can be heard very far away.[8]

As nobody is willing anymore to invest into AM towers and equipment, only one type of antenna will be described for this band – the one you could use to receive AM signal via your FM/AM tuner.

Small Loop Antenna

It is very simple antenna, just a simple coil. The mount must be from an insulator. Two designs are the most common:

- 12 windings, height 10cm, width 12cm.
- 5 windings, height 12cm, width 16cm.

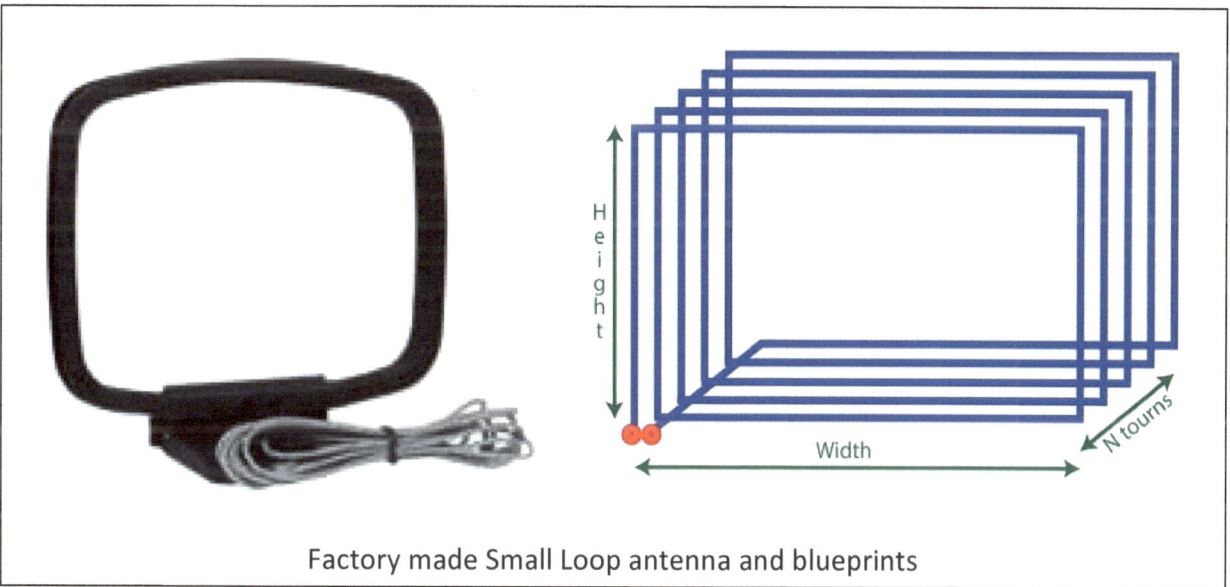

Factory made Small Loop antenna and blueprints

---

[8] Some of the reasons why it is less and less used band.

FM

Intro

FM band lies between 88MHz and 108MHz. Most often used antennas in this band are dipole or some variant of it, and higher gain is achieved by vertical stacking. Polarization used is vertical or circular.

Dipole

Indoor wire dipole antenna 72 Ω

For receiving only, we will choose 100MHz as our target frequency as it will cover all the band.

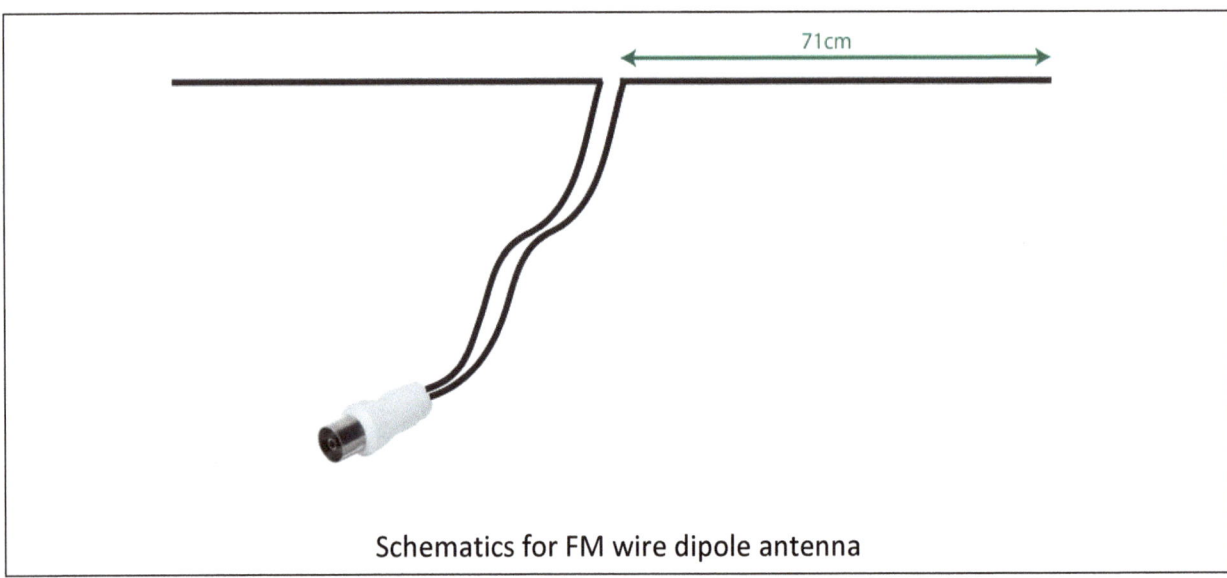

Schematics for FM wire dipole antenna

Materials:

1. Any twin-lead cable,
2. RF connector.

Construction

1. Cut the twin lead cable to the length of 142cm.
2. Use hot glue gun and glue over the half of the twin-lead.
3. Split the twin lead cable up to the glued part, so that you get two wires each 71cm long.
4. On the other side attach RF connector.

Indoor wire folded dipole antenna 300Ω

It is a variant of the previous antenna, with slightly better bandwidth. The choice should be based on the impedance of the FM receiver.

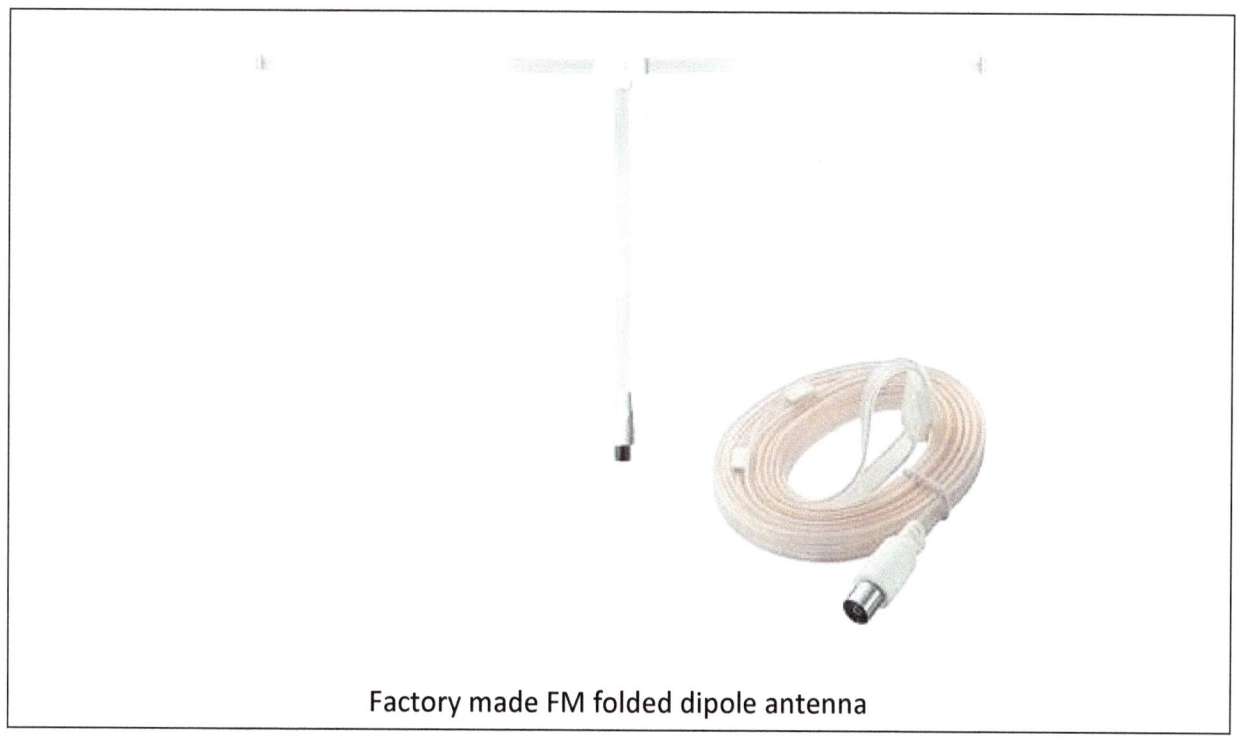

Factory made FM folded dipole antenna

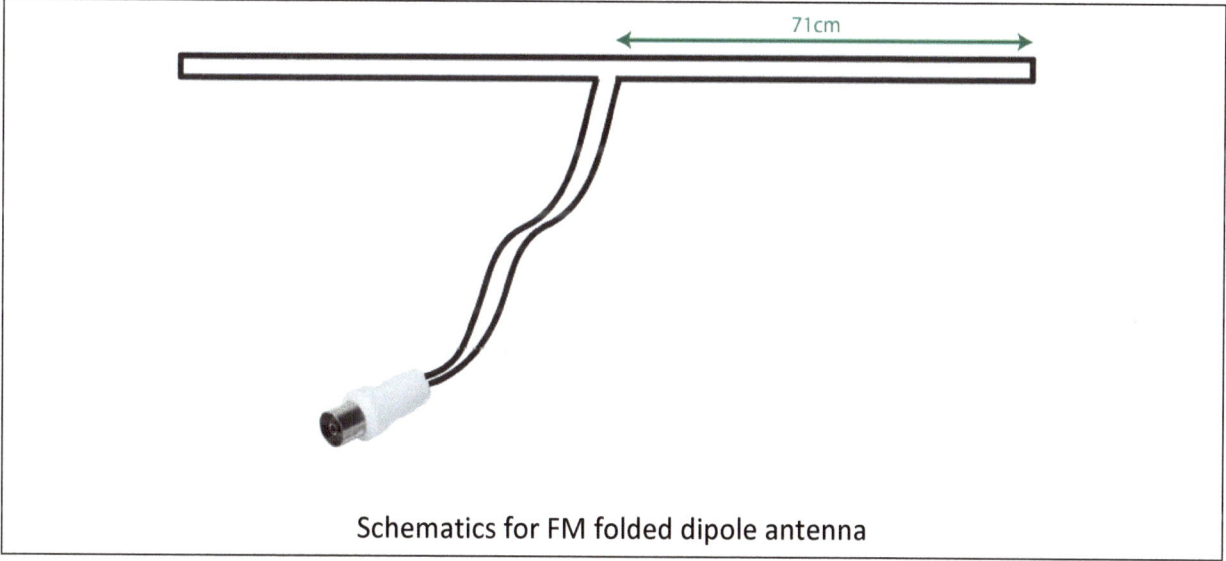

71cm

Schematics for FM folded dipole antenna

Materials:

1. Any twin-lead cable,
2. RF connector.

Construction

1. Cut the twin lead cable into two pieces, one that is 142cm long and the other 71cm long.

2. Cut one lead of the longer cable in half, leaving the other intact.
3. Remove insulation from the ends of each side of both cables.
4. From the longer cable, twist the ends and weld them together.
5. Twist the ends of the shorter cable with the middle of the longer cable and weld them together.
6. Cut the excessive length and insulate welded parts.
7. On the other side of the shorter cable attach RF connector.

Wire dipole 72 Ω

This is the simplest outdoor FM dipole. It is very simple to make, but the pipe version is recommended because it has greater bandwidth. This version will not work well at the edges of the band.

Materials:

1. PVC pipe with 22mm diameter,
2. Square metal piece,
3. Two U shape screws,
4. PVC pipe holders,
5. Wire clamps,
6. Steel or brass wire,
7. RG-6 coaxial cable.

Construction

1. Cut the PVC pipe to the length 0.5m to 1m,
2. Drill 6 holes in the square metal piece like on diagram,
3. Drill 2 holes in the PVC pipe,
4. Cut one half of the top of the PVC pipe,
5. Attach wire clamps and hot glue it,
6. Cut the hole into the pipe to pass the coaxial cable,
7. Connect the ends of the coaxial cable to the wire clamps,
8. Add 2 pieces of 72cm straight wire to the wire clamps, and tighten the screws,
9. Use hot glue gun to seal everything,
10. Assemble the holder and connect it with the pipe.

Dipole from aluminum pipes

This is the best choice for outdoor dipole if the FM receiver has impedance 75Ω.

Materials:

1. Two aluminum pipes with diameter 6mm,

2. PVC pipe with 22mm diameter,
3. Square metal piece,
4. Two U shape screws,
5. PVC pipe holders,
6. T-shape PVC pipe connector,
7. Two 5mm screws and small metal strips,
8. RG-6 coaxial cable.

Construction

1. Do steps 1-3 from the previous antenna,
2. Make holders for pipes from screws ⌀5mm (lower-right photo),
3. Put RG-6 cable through the plastic pipe and connect hot-lead and ground to each screw,
4. Isolate them with hot-glue,
5. Cut the ⌀6mm aluminum pipes to the measure,
6. Cut the ⌀8mm aluminum pipe to the length of a few centimeters (we need two of them),
7. Put the ⌀8mm pipe pieces over the ends of the ⌀6mm pipes (lower-right photo),
8. Drill through the axes inside ⌀6 pipes with ⌀4.5mm drill (inside it is ⌀4, we want it a bit wider),
9. Make ⌀5mm threads inside each pipe,
10. Close the other ends of the pipe with hot-glue,
11. Assemble – screw pipes onto the holder.

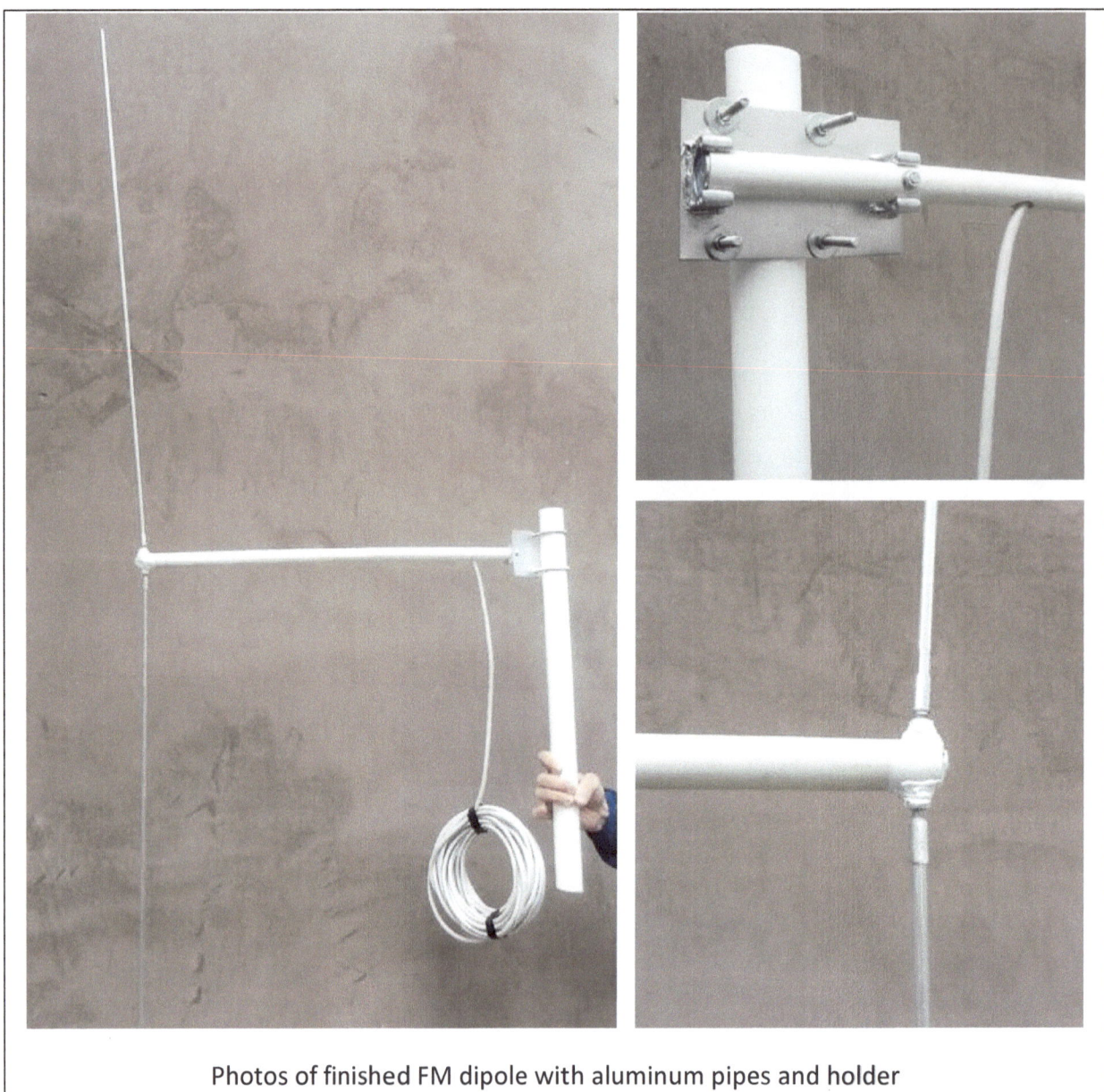

Photos of finished FM dipole with aluminum pipes and holder

Note: This antenna has been tested to withstand strong winds for years, but it is not guarantee how strong storms can it withstand. Folded dipole is much stronger.

Folded dipole from aluminum pipes

This antenna is recommended when FM receiver has 300Ω impedance. It is the best receiving dipole, because it has the greatest bandwidth.

The instructions for this antenna are the same as for the next antenna, but we can use thinner pipes.

Transmitting folded dipole antenna

The construction is evident from the image. It is important that folded dipole is isolated from the metal boom. PVC boom can be used as well or the plastic or silicone standoffs on metal boom.

The thicker the pipe (the larger cross-section area), the more power it can handle. The diameter of the pipe is proportional to the bandwidth.

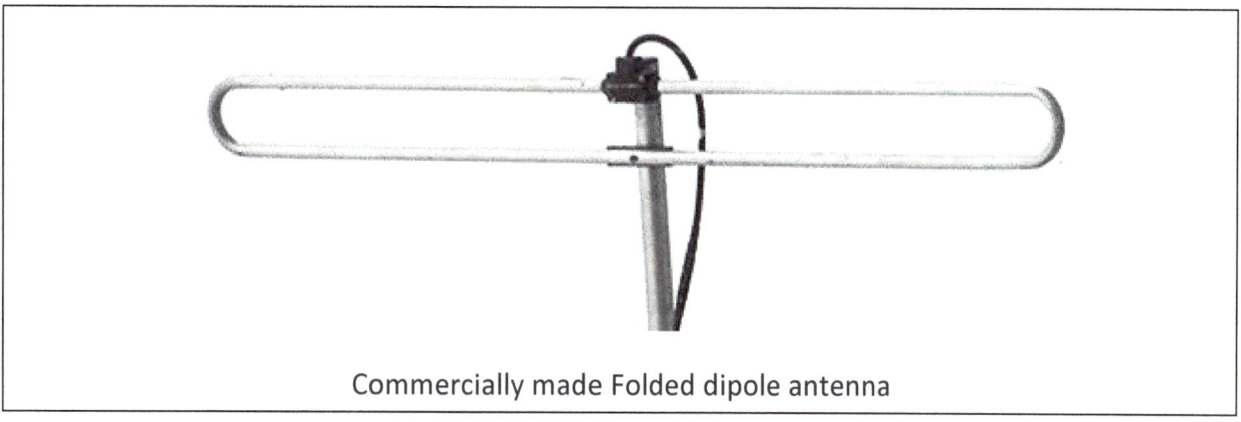

Commercially made Folded dipole antenna

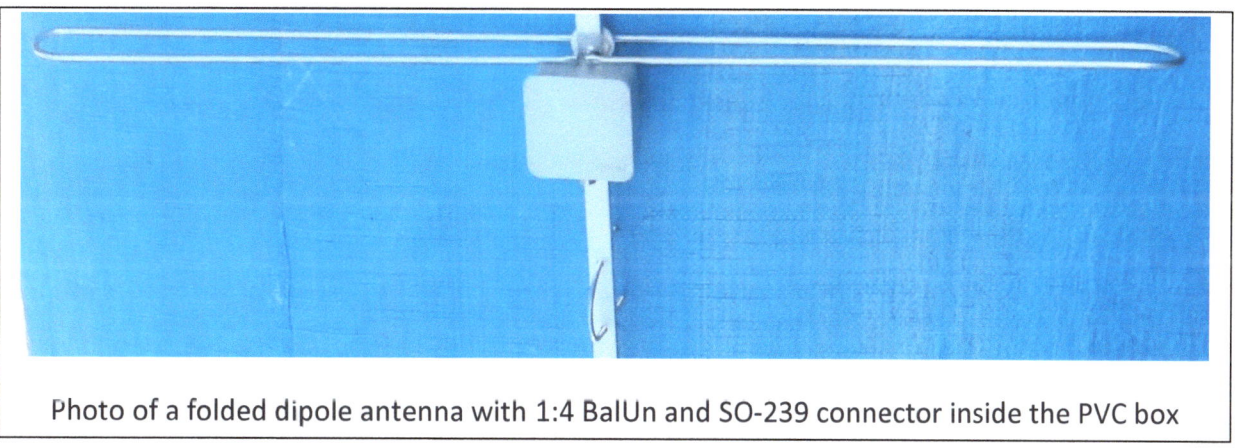

Photo of a folded dipole antenna with 1:4 BalUn and SO-239 connector inside the PVC box

High-Power transmission

Special care should be taken so that the connectors, cable and BalUn can handle the power. RG-58 is often not suitable for higher power signals. PL and SO239 are good choice of connectors for this application. If BalUn is made from a coaxial cable, not transformer, it should be made with at least RG-213 cable.

Transmitting Ground Plane (GP) antenna at 100MHz
- Radial length = 75mm
- Vertical length = 71mm
- Radial material: aluminum pipe with diameter 6mm.

For construction details return to "Ground Plane (GP)" section.

Transmitting and receiving Yagi 3 antenna

For more on Yagi design, see the "Yagi" section.

Gain of ideally made antenna is 6.9dBi.

- Diameter of dipole bend = 60mm,
- Dipole gap at feed point = 20mm,
- Cross section of square boom = 40mm,
- Dipole is made from pipe with 10mm diameter,
- Directors are pipes with diameter 8mm, bonded through metal boom,
- Folded dipole length = 1447mm, total rod length = 2042mm, center of rod is at 1471mm. Parameters: BC = 693, HI = 683, HA = 731, HB = 778.

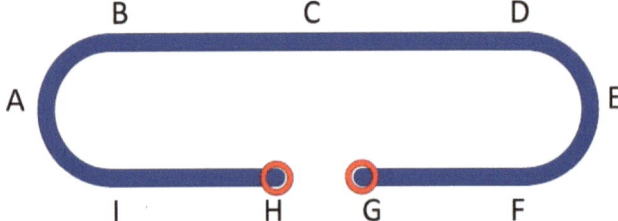

- BalUn from 50Ω coaxial wire length = 989mm,
- Reflector length = 1482mm at boom position 30mm. Insert it to 721mm,
- Dipole is spaced from reflector by 600mm, at boom position 630mm. Insert it to 703.5mm.
- Director length = 1363mm, spaced from dipole by 225mm at boom position 854mm. Insert it to 661.5mm.
- Boom length is 884mm.

TV

Intro

Polarization used in TV broadcasting is mostly horizontal. The standard antennas that are being used for receiving TV signal are Log and Yagi. They have been improved to their maximum and are manufactured in great numbers and cheaply, thus they are not covered in this book. Only two TV antennas are covered: Fractal and Gray-Hoverman antenna. They have several advantages, one of which is that they are great for receiving DTV. We have tested Gray-Hoverman antenna versus the Log, Panel antenna with amplifier, and Yagi for receiving the DTV signal, without optical sight from the transmitter, for DVB-T2 standard. Home-made Gray-Hoverman antenna outperformed them all in the given circumstances.

Fractal Antenna

Fractal antennas are the latest and the most advanced designs. It uses self-similar designs and, as a result, one gets reduced antenna size (in comparison with classical antenna types for the same gain, or better gain per size, if you like) and better bandwidth. They can be used for all range of frequencies, but here we will explore the example that can be used for the whole range of TV broadcasting frequencies which is approximately 200-800MHz. The emphasis is put on the VHF, because UHF stations are usually more powerful and there are more VHF stations. Thus you would get equally good reception from VHF and UHF. Although any fractal can be used, the easiest to make is the Koch curve.

Simulation

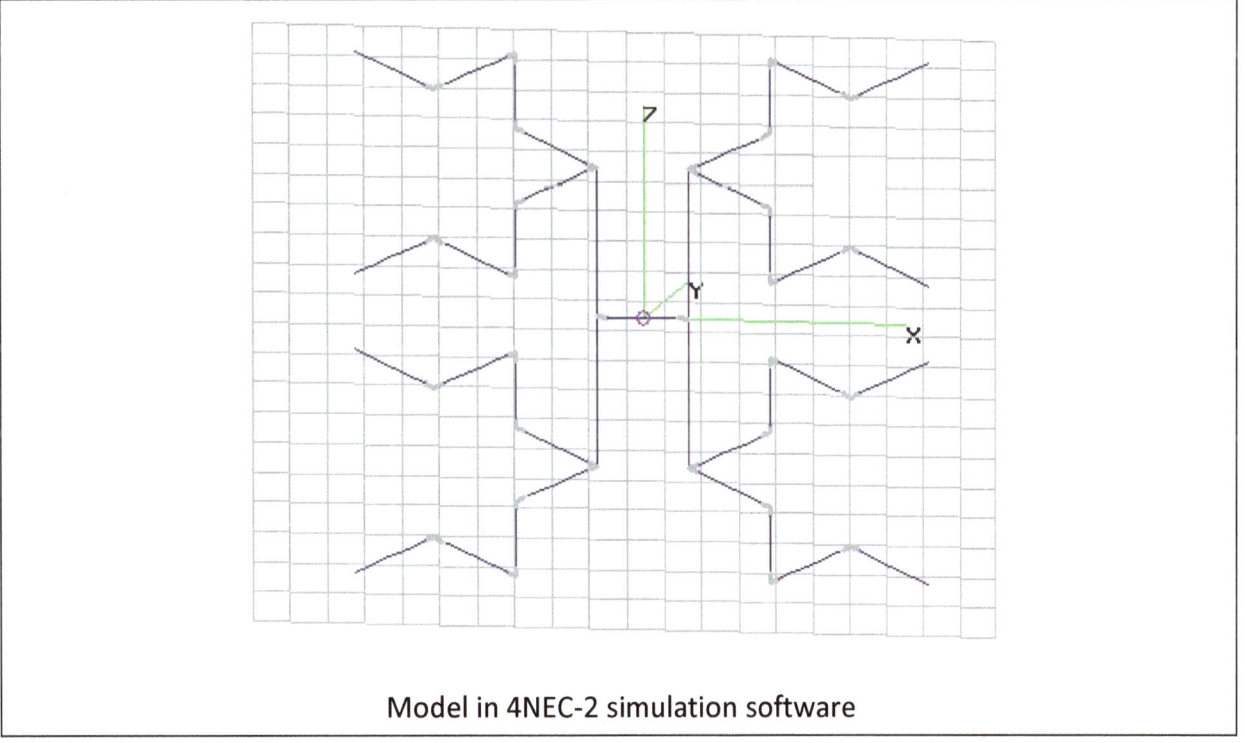

Model in 4NEC-2 simulation software

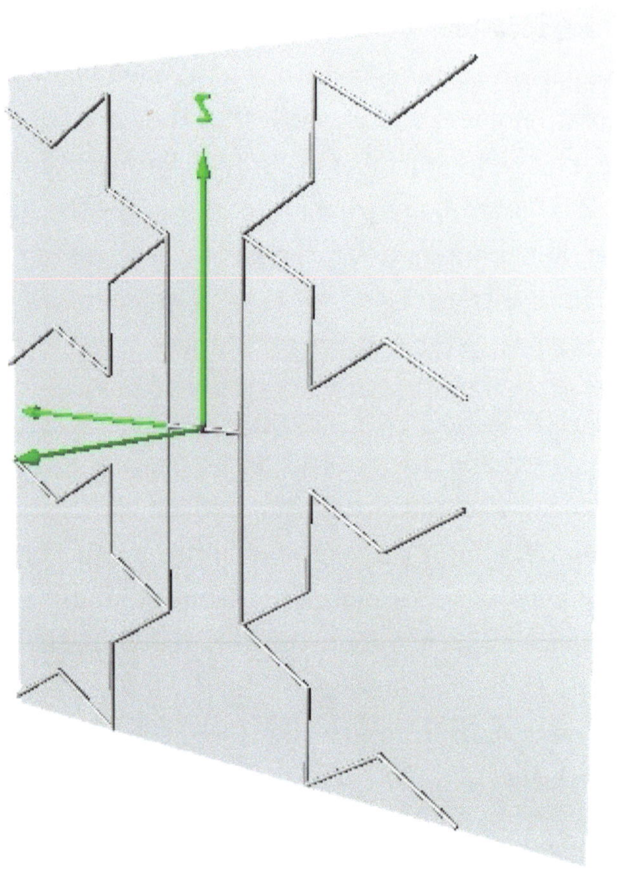

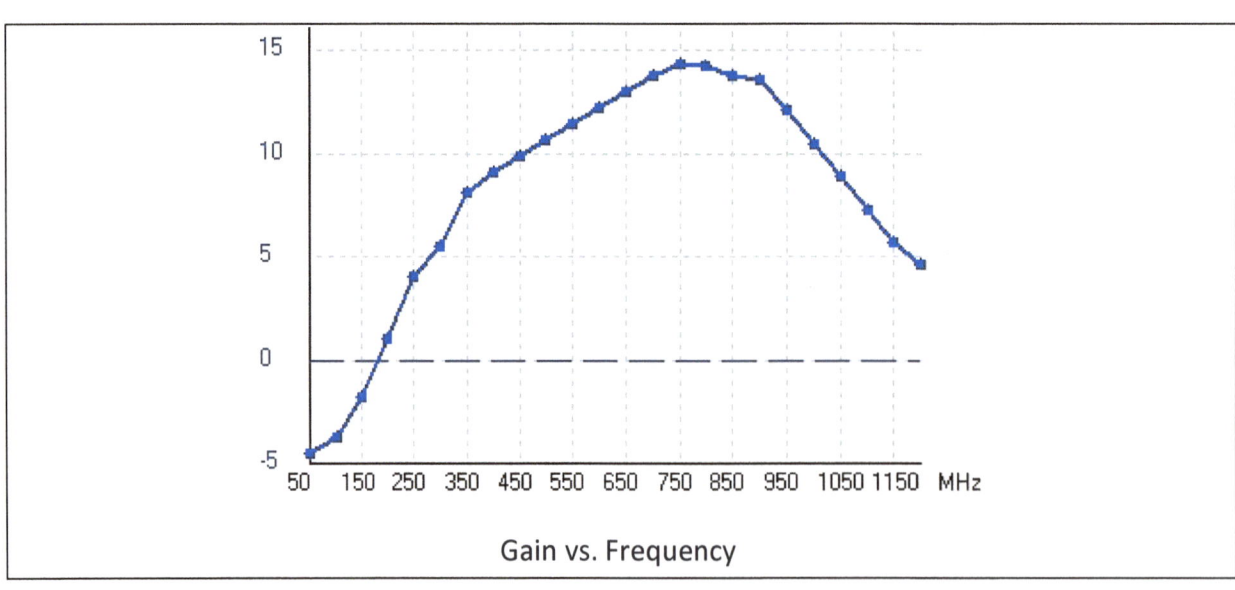

3D Model in 4NEC-2

Gain vs. Frequency

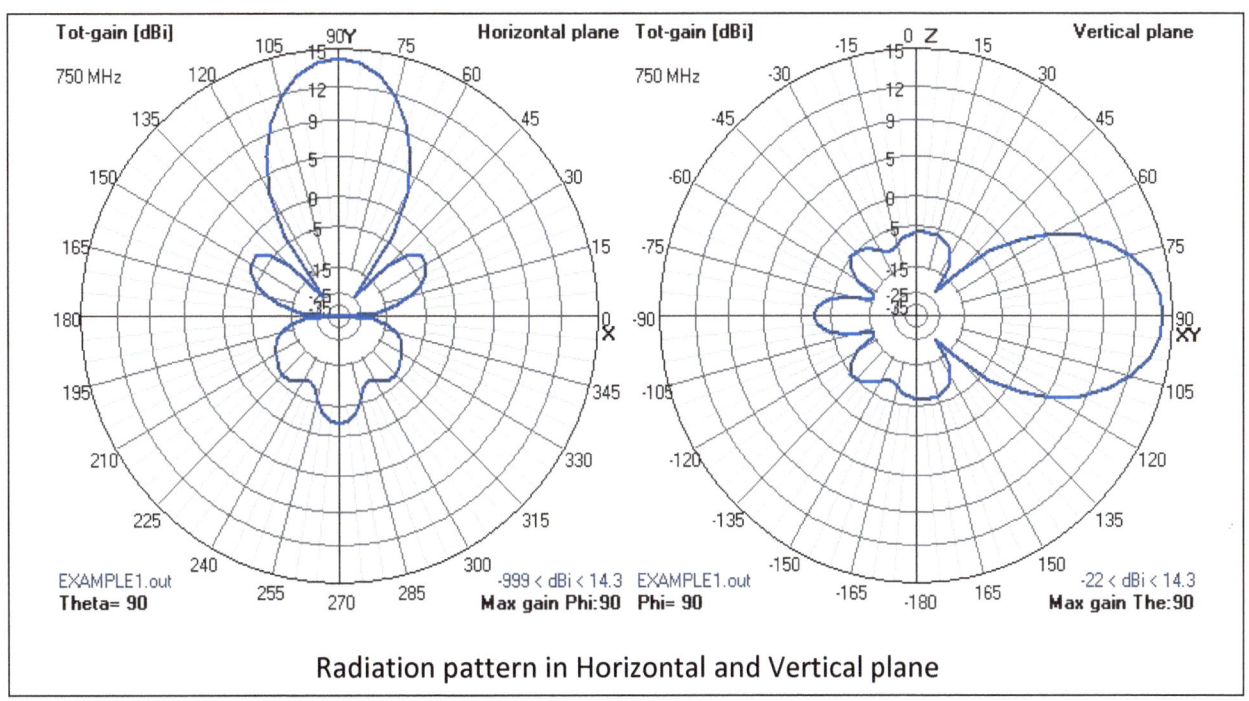

Radiation pattern in Horizontal and Vertical plane

Construction

Radiating element is best made from brass, at least 3mm thick. The pattern follows the six point star.

The reflector is best made from Wireframe of dimensions 280 mm × 56 mm. Wireframe should have enough density, in practice 20mm × 20mm is good enough choice.

The distance between radiating element and the reflector should be 70 mm.

Standoffs should be made from non-conductive element – wood or PVC.

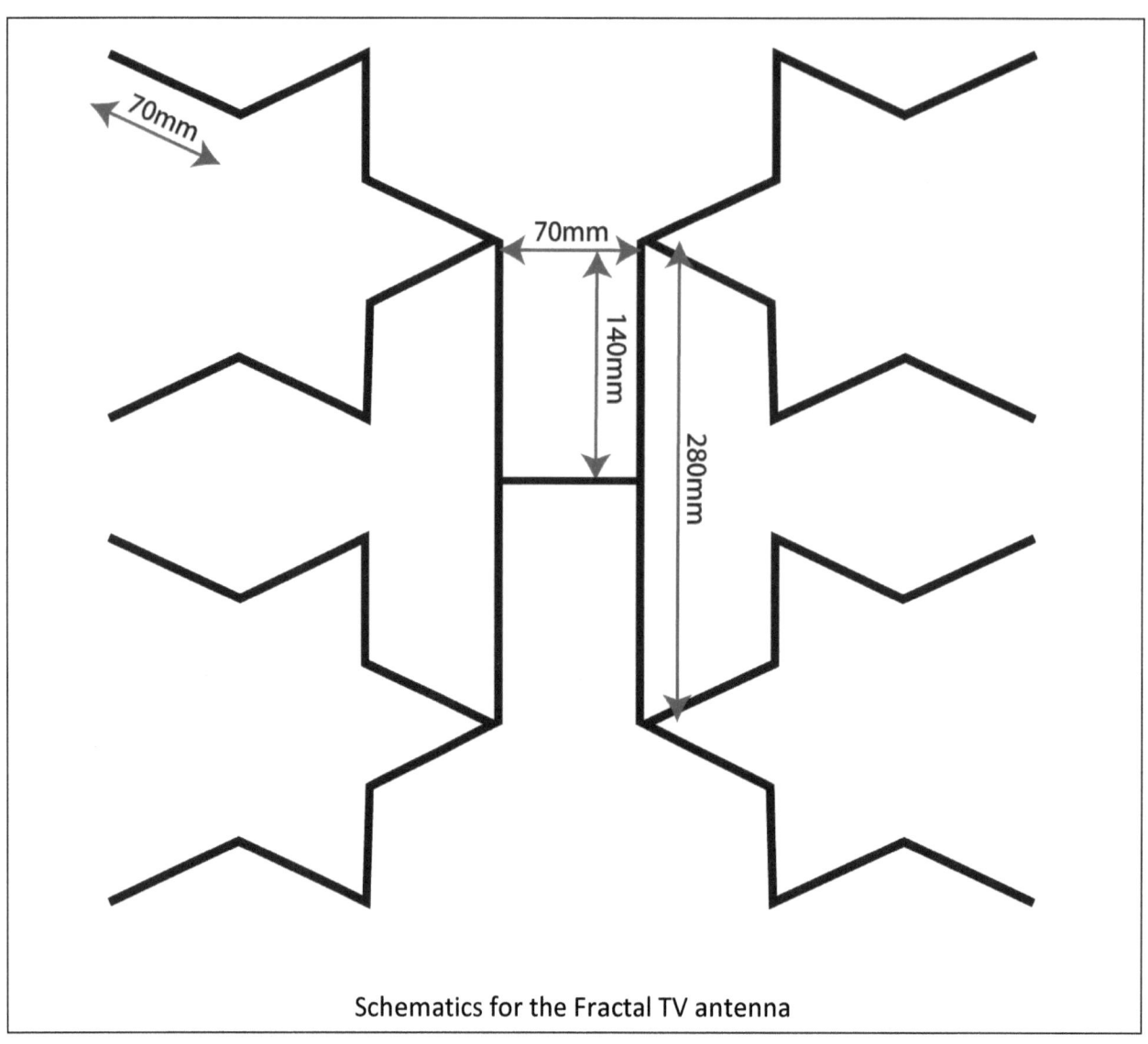

Schematics for the Fractal TV antenna

Gray-Hoverman antenna

Very good antenna for both analogue TV and DTV, especially DTV as it outperforms cheap Yagi's, Log, panel and other commercial antennas. It is designed for the UHF region, but above channel 54, the gain drops significantly. The antenna has 14dBi at its resonant frequency, and does not drop under 7dBi, with an average of 12dBi taken over all the channels.

Simulation

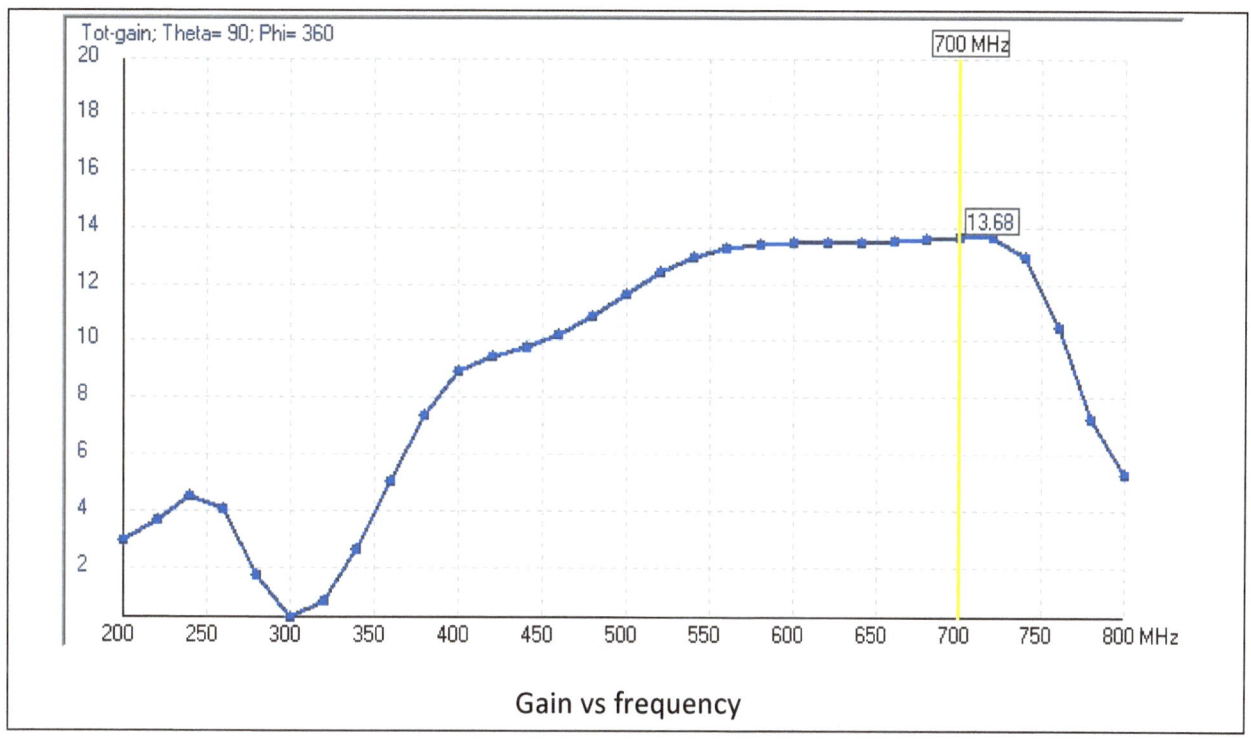

Gain vs frequency

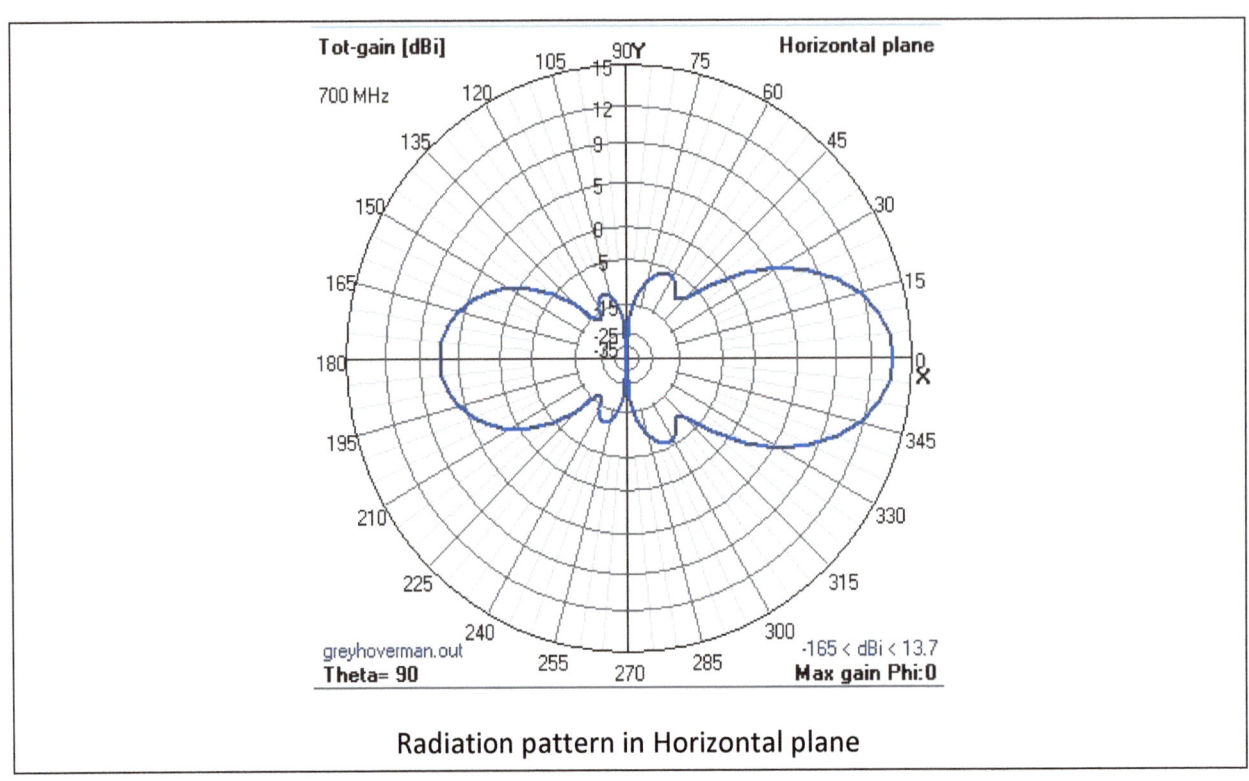

Radiation pattern in Horizontal plane

Reflector can be made from net wire or with rods with similar performance. It is simpler to make it with rods. All rods can be made from any diameter and any metal, although the best results are achieved with higher diameter and more expensive materials, like brass. Copper is not recommended because it is very soft, and it would get bent by wind quite easily.

Construction

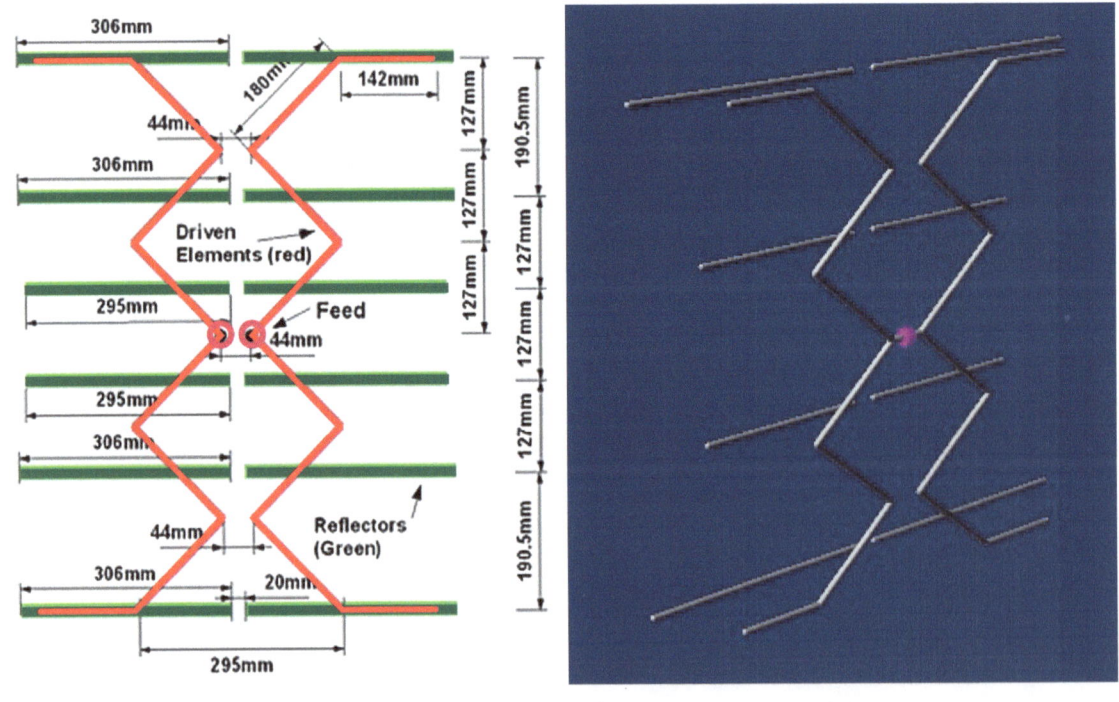

4nec2

Save this text as a grayhoverman.nec file and open it in 4nec2 simulation software:

```
CM - Gray-Hoverman
CE
GW 19 15 4 -1 9.5 4 -6.25 4.75 0.04040404
GW 20 15 4 -6.25 4.75 4 -1 0 0.04040404
GW 21 15 4 -1 0 4 -6.25 -4.75 0.04040404
GW 22 15 4 -6.25 -4.75 4 -1 -9.5 0.04040404
GW 23 15 4 -1 -9.5 4 -6.25 -14.25 0.04040404
GW 24 11 4 -6.25 -14.25 4 -11.25 -14.25 0.04040404
GW 25 15 4 -1 9.5 4 -6.25 14.25 0.04040404
GW 26 11 4 -6.25 14.25 4 -11.25 14.25 0.04040404
GW 27 15 4 1 9.5 4 6.25 4.75 0.04040404
GW 28 15 4 6.25 4.75 4 1 0 0.04040404
GW 29 15 4 1 0 4 6.25 -4.75 0.04040404
GW 30 15 4 6.25 -4.75 4 1 -9.5 0.04040404
GW 31 15 4 1 -9.5 4 6.25 -14.25 0.04040404
GW 32 11 4 6.25 -14.25 4 11.25 -14.25 0.04040404
GW 33 15 4 1 9.5 4 6.25 14.25 0.04040404
GW 34 11 4 6.25 14.25 4 11.25 14.25 0.04040404
GW 35 5 4 -1 0 4 1 0 0.0285341
GW 36 29 0 -0.75 -14 0 -14.75 -14 0.1875
GW 37 21 0 -0.75 -5 0 -10.75 -5 0.1875
GW 38 21 0 -0.75 5 0 -10.75 5 0.1875
GW 39 29 0 -0.75 14 0 -14.75 14 0.1875
GW 40 29 0 0.75 14 0 14.75 14 0.1875
GW 41 21 0 0.75 5 0 10.75 5 0.1875
GW 42 21 0 0.75 -5 0 10.75 -5 0.1875
GW 43 29 0 0.75 -14 0 14.75 -14 0.1875
GS 0 0 0.0254 ' All in inches.
GE 0
EK
EX 0 35 3 0 1 0
GN -1
' FR 0 57 0 0 473 6 ' from 300Ω
' RP 0 1 73 1510 90. 0. 1 5 0. 0. ' from 300Ω
' FR Freq Sweep choices in order of increasing calculation time (fm holl_ands):
```

```
' FR 0 0 0 0 470 0 ' Fixed Freq

FR 0 29 0 0 470 12 ' Freq Sweep 470-806 every 12 MHz - OLD UHF BAND

' FR 0 34 0 0 410 12 ' Freq Sweep 410-806 every 12 MHz - Even Wider Sweep

' FR 0 39 0 0 470 6 ' Freq Sweep 470-698 every 6 MHz - PREFERRED FOR UHF

' FR 0 77 0 0 470 3 ' Freq Sweep 470-698 every 3 MHz

' FR 0 153 0 0 470 1.5 ' Freq Sweep 470-698 every 1.5 MHz

' FR 0 71 0 0 300 10 ' Freq Sweep 300-1000 every 10 MHz - WIDEBAND SWEEP

' FR Hi-VHF choices:

' FR 0 15 0 0 174 3 ' Freq Sweep 174-216 every 3 MHz

' FR 0 29 0 0 174 1.5 ' Freq Sweep 174-216 every 1.5 MHz - PREFERRED

' FR 0 45 0 0 162 1.5 ' Freq Sweep 162-228 every 1.5 MHz - Add +/- 12 MHz BW

' FR 0 43 0 0 174 1 ' Freq Sweep 174-216 every 1 MHz - Hi-Rez

' FR 0 23 0 0 198 1 ' Freq Sweep 198-220 every 1 MHz - Hi-Rez - Ch13 SPECIAL

' FR 0 26 0 0 150 6 ' Freq Sweep 150-300 every 6 MHz - WIDEBAND SWEEP

' FR Lo-VHF choices:

' FR 0 19 0 0 54 3 ' Frequency Sweep every 3 MHz for Ch2-6 + FM

' FR 0 35 0 0 54 1 ' Frequency Sweep every 1 MHz for Ch2-6

' FR 0 36 0 0 75 1 ' Frequency Sweep every 1 MHz for Ch5 + Ch6 + FM

' FR 0 28 0 0 54 6 ' Wide Freq Sweep every 6 MHz for Ch2-13

' FR 0 64 0 0 54 12 ' Super Wide Freq Sweep 54-810 every 12 MHz

' FR 0 127 0 0 54 6 ' Super Wide Freq Sweep 54-810 every 6 MHz

' RP choices in order of increasing calculation time:

' RP 0 1 1 1510 90 90 1 1 0 0 ' 1D Gain toward 0-deg Azimuth - SIDE GAIN

' RP 0 1 1 1510 90 0 1 1 0 0 ' 1D Gain toward 90-deg Azimuth - FORWARD GAIN

' RP 0 1 1 1510 90 180 1 1 0 0 ' 1D Gain toward 270-deg Azimuth - REVERSE GAIN

' RP 0 1 37 1510 90 0 1 5 0 0 ' 2D (Left only) Azimuthal Gain Slice

RP 0 1 73 1510 90 0 1 5 0 0' 2D Azimuthal Gain Slice - PREFERRED

' RP 0 73 1 1510 90 0 5 1 0 0 ' 2D Elevation Gain Slice

' RP 0 73 73 1510 90 0 5 5 0 0 ' 3D Lower Hemisphere reveals antenna (Fixed Freq)

' RP 0 285 73 1510 90 0 5 5 0 0 ' 3D Full Coverage obscures antenna (Fixed Freq)

EN
```

HAM (Radio Amateur) bands

Multiband antennas

Universal Dipole for CW and SBB

This is Inverted V antenna that can be used for both CW and SBB. SBB part is resonant at 3735 kHz, for CW at 3535 kHz.

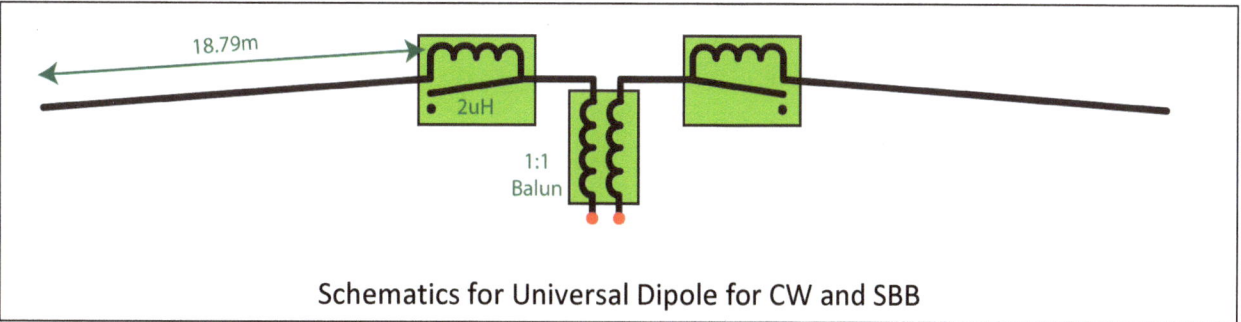

Schematics for Universal Dipole for CW and SBB

12V relays are used to switch between SSB and CW, two coils are used to lower the frequency.

Wire used for one arm of Inverted V is 18.79m, total length of antenna is 37.58m, with wire diameter approximately 1mm.

Coil construction: cut two wires of diameter 1.6mm and length 933mm and wound them to cylinder of diameter 32mm. After winding, remove cylinder and spread windings apart so that distance between them is 1-2mm. By spreading the windings further or closer apart you can fine-tune the CW resonant frequency.

Operation: When there is no voltage across relays, antenna works in SBB mode. When 12V is applied across relays, inductive elements start to work and antenna works in CW mode. This means that you need 12V power source with a switch as well. Relays are best fed through separate twin-lead cable.

BalUn: if the compact solution is needed, instead of the Air core RF choke, we can use torus ferrite core. Details about construction are in "BalUn" section of the book.

Multiband dipole for 40m, 80m and 160m

This antenna covers 65kHz for 160m band, 75kHz for 80m band and the entire 40m band with SWR under 2:1. The tradeoff is limited bandwidth.

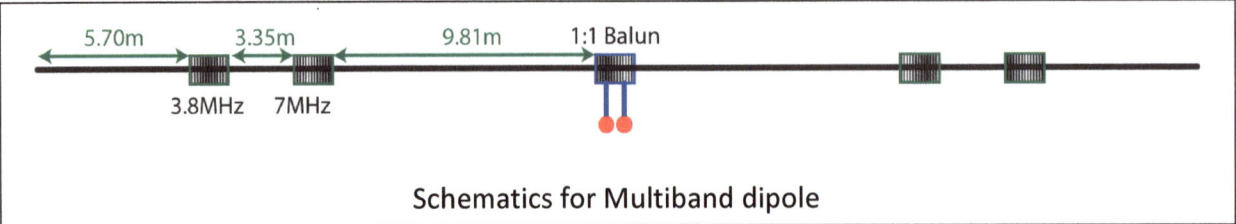

Schematics for Multiband dipole

The coils used are traps for the signal – they act as a notch filters[9].

The shortest segment (9.81m) is for 40m, the next segment (9.81m + 3.35m) is for 80m and the whole antenna (9.81m + 3.35m + 5.70m) is for 160m.

We get wire for the coils when we strip the shield from the RG-58 coaxial cable. We thus get only the core – the hot wire, and the first isolation. The coil forms are cut from the PVC pipe.

For the 7MHz trap, we need PVC pipe with 6cm outer diameter. The windings are in two levels: the number of inner windings is 12 $1/_3$ turns, and 11 ¼ of outer turns.

The 3.8MHz trap form is made from 3.5" outer diameter pipe. It has 14.3 turns for the inner winding and 13.4 turns for the outer winding.

Use electrical tape to keep the turns in place. Try to make them as tight and closely spaced as possible. Drill the hole in PVC pipe where the winding will enter and exit and secure that hole with hot glue so the coil does not unwind.

---

[9] The windings have capacitance as well.

Extended AWX for 2m and 70cm bands

AWX stands for All Wave, and it is shaped like an X. Gain is higher than for dipole antenna, and it can be used in very wide frequency range.

One quarter of a wavelength for 145MHz is equal to 984mm, for 435MHz 328mm. The distance at a feed point should be as small as possible.

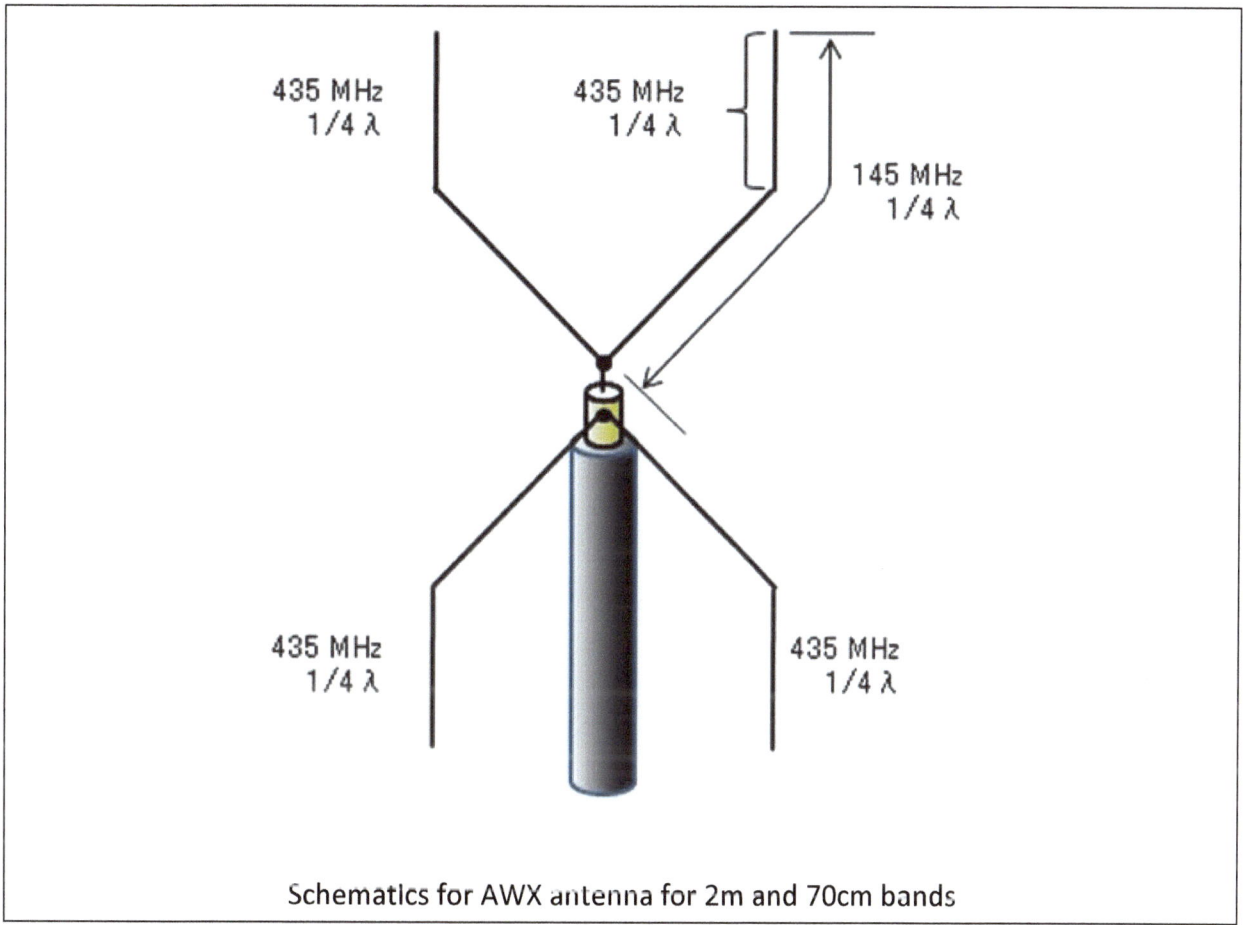

Schematics for AWX antenna for 2m and 70cm bands

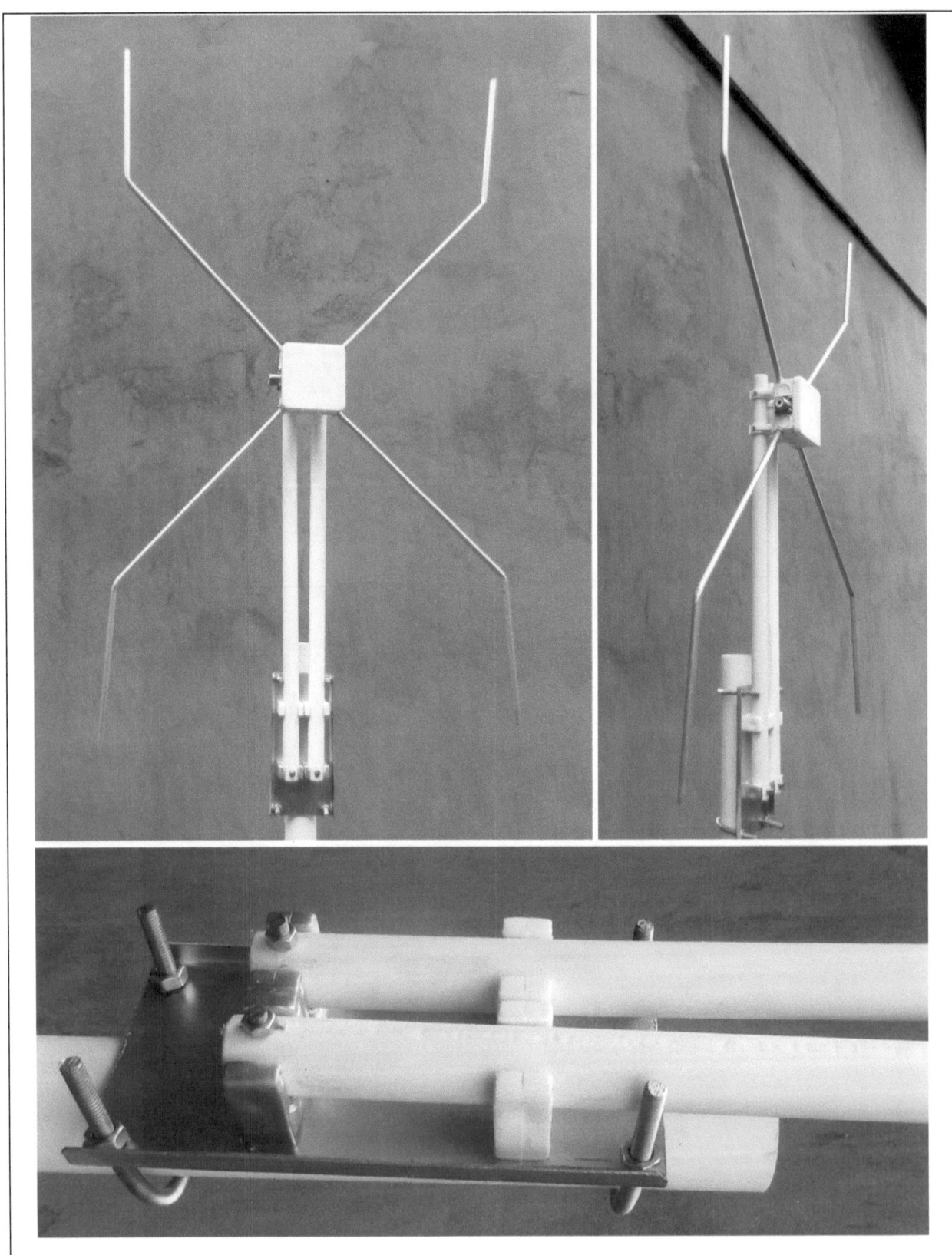

Top-left: frontal view, top-right: side view, bottom: holder

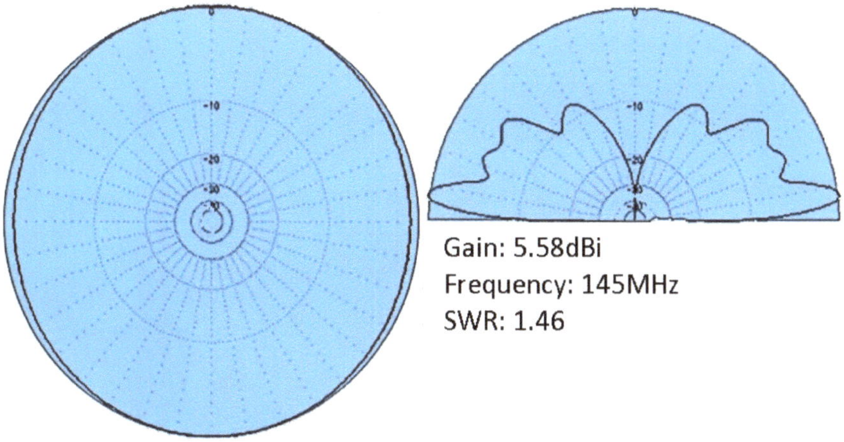

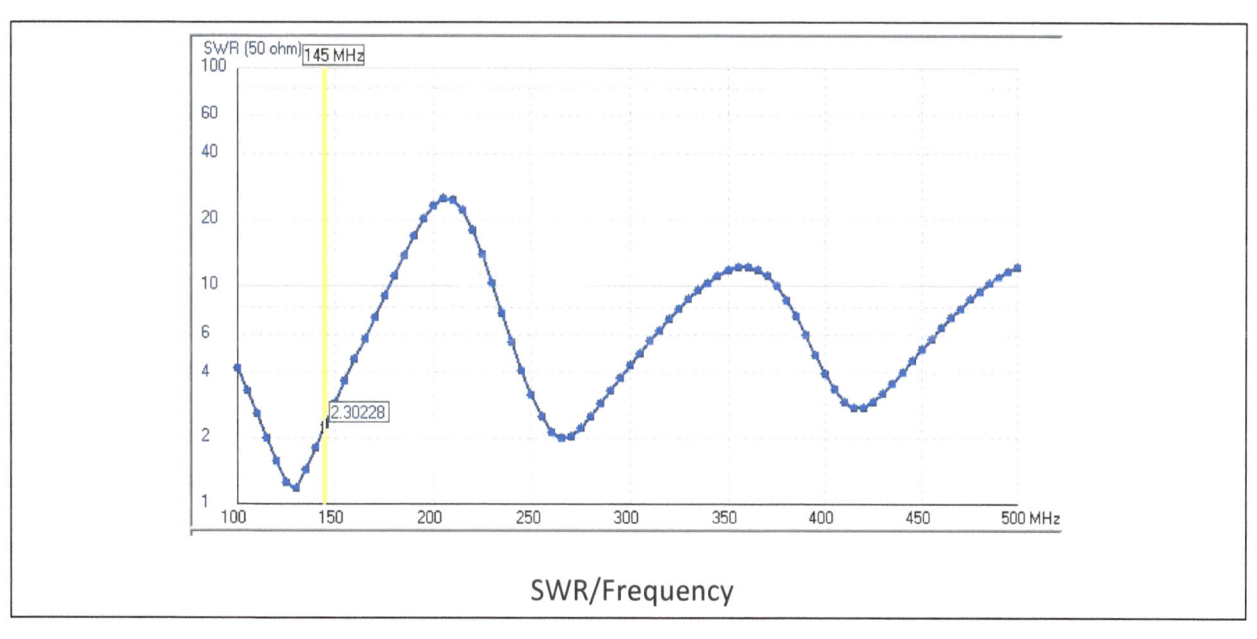

Gain: 5.58dBi
Frequency: 145MHz
SWR: 1.46

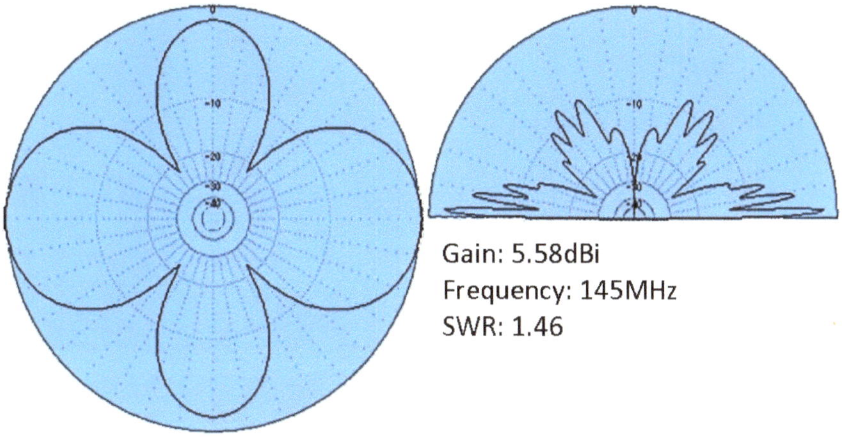

Gain: 5.58dBi
Frequency: 145MHz
SWR: 1.46

Horizontal radiation is on the left, vertical on the right

SWR/Frequency

Dual band GP for 2m and 70cm band

Dual band GP is like two single band antennas have merged, one rotated by 45° arround the z-axis. Connections are the same for the two antennas: radiator is connected to the same hot-end of the connector, and grounds are connected to the same mass.

This way there is no limit to how many bands antenna can be made. In practice, though, most common are two and three band GP antennas.

For measurements, see the "GP" section.

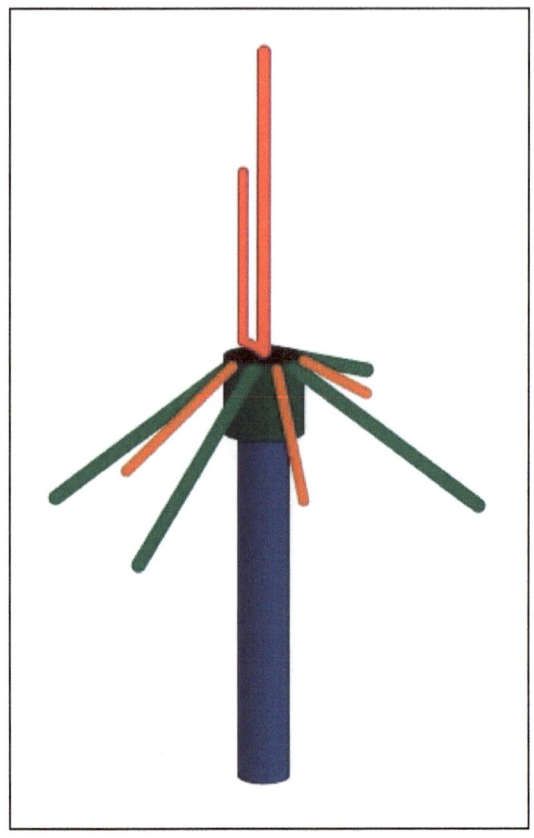

Photo of a commercial dual-band antenna used for long-distance Senao phones

160m (1800-2000 kHz)

Dipole, Inverted V and Delta Loop

For more information, see the Dipole section.

- Dipole total length = 75m, one arm length = 37.5m
- Inverted V total length = 71.3m, one arm length = 36.65m
- Delta Loop total length = 157.2m, one side length = 52.4m

Continuous Loaded

For more information, see the Vertical monopole with Continuous Loaded coil antenna section.

Wind from top to bottom, leaving straight wire at the bottom from the end of the coil to the end of the mast.

- Coil Length = 2116mm
- Number of turns = 753
- Wire diameter = 2.59mm
- Diameter of coil = 51mm
- Mast length = 2190mm
- Loss = 21.6db

80m (3500-4000 kHz)

Dipole, Inverted V and Delta Loop

For more information, see the Dipole section.

- Dipole total length = 40.8m, one arm length = 20.4m
- Inverted V total length = 38.7m, one arm length = 19.35m
- Delta Loop total length = 78.17m, one side length = 26.06m

Continuous Loaded

For more information, see the Vertical monopole with Continuous Loaded coil antenna section.

Wind from top to bottom, leaving straight wire at the bottom from the end of the coil to the end of the mast.

- Coil Length = 2116mm
- Number of turns = 824
- Wire diameter = 2.05mm
- Diameter of coil = 25mm

- Mast length = 2190mm
- Loss = 15db

60m (5MHz)

Dipole, Inverted V and Delta Loop

For more information, see the Dipole section.

- Dipole total length = 28.5m, one arm length = 14.25m
- Inverted V total length = 27.1m, one arm length = 13.55m
- Delta Loop total length = 61.26m, one side length = 20.42m

40m (7.0-7.3 MHz)

Dipole, Inverted V and Delta Loop

For more information, see the Dipole section.

- Dipole total length = 20.4m, one arm length = 10.2m
- Inverted V total length = 19.4m, one arm length = 9.7m
- Delta Loop total length = 40.79m, one side length = 13.60m

Continuous Loaded vertical antenna

For more information, see the Vertical monopole with Continuous Loaded coil antenna section.

Wind from top to bottom, leaving straight wire at the bottom from the end of the coil to the end of the mast.

- Coil Length = 2116mm
- Number of turns = 405
- Wire diameter = 2.59mm
- Diameter of coil = 25mm
- Mast length = 2190mm
- Loss = 5.5db

30m (10.10-10.15 MHz)

Dipole, Inverted V and Delta Loop

For more information, see the Dipole section.

- Dipole total length = 14.3m, one arm length = 7.15m

- Inverted V total length = 13.5m, one arm length = 6.75m
- Delta Loop total length = 28.68m, one side length = 9.56m

Yagi 2

For more information, see the Yagi 2 section.

- Boom Length: 3m,
- Dipole Length: 14.92m
- Director Length: 14.04m
- Impedance: 28 Ω

20m (14.00-14.35 MHz)

Dipole, Inverted V and Delta Loop

For more information, see the Dipole section.

- Dipole total length = 10.2m, one arm length = 5.1m
- Inverted V total length = 9.7m, one arm length = 4.85m
- Delta Loop total length = 20.41m, one side length = 6.80m

Continuous loaded vertical antenna

For more information, see the Vertical monopole with Continuous Loaded coil antenna section.

Wind from top to bottom, leaving straight wire at the bottom from the end of the coil to the end of the mast.

- Coil length = 2116mm
- Number of turns = 191
- Wire diameter = 2.59mm
- Diameter of coil = 25mm
- Mast length = 2190mm
- Loss = 1db

Yagi 2

For more information, see the Yagi 2 section.

- Boom Length: 2.3m
- Dipole Length: 10.26m
- Reflector Length: 11.08m
- Impedance: 28 Ω

17m (18.068-18.168 MHz)

Dipole, Inverted V and Delta Loop

For more information, see the Dipole section.

- Dipole total length = 7.9m, one arm length = 3.95m
- Inverted V total length = 7.5m, one arm length = 3.75m
- Delta Loop total length = 17.01m, one side length = 5.67m

Yagi 2

For more information, see the Yagi 2 section.

- Boom Length: 1.7m
- Dipole Length: 8.32m
- Director Length: 7.80m
- Impedance: 28 Ω

15m (21.00-21.45 MHz)

Dipole, Inverted V and Delta Loop

For more information, see the Dipole section.

- Dipole total length = 6.8m, one arm length = 3.4m
- Inverted V total length = 6.5m, one arm length = 3.25m
- Delta Loop total length = 13.65m, one side length = 4.55m

Yagi 2

For more information, see the Yagi 2 section.

- Boom Length: 1.61m
- Dipole Length: 6387m
- Reflector Length: 7.22m
- Impedance: 28 Ω

12m (24.89-24.99 MHz)

Dipole, Inverted V and Delta Loop

For more information, see the Dipole section.

- Dipole total length = 5.94m, one arm length = 2.97m
- Inverted V total length = 5.64m, one arm length = 2.82m
- Delta Loop total length = 11.66m, one side length = 3.88m

Yagi 2

For more information, see the Yagi 2 section.

28 Ω

- Boom Length: 1.18m
- Dipole Length: 6.05m
- Director Length: 5.60m

12.5 Ω

- Boom Length: 6.7m
- Dipole Length: 6.09m
- Director Length: 5.78m

11m (27MHz) – CB

Dipole, Inverted V and Delta Loop

For more information, see the Dipole section.

- Dipole total length = 5.28m, one arm length = 2.64m
- Inverted V total length = 5.02m, one arm length = 2.51m
- Delta Loop total length = 11.35m, one side length = 3.78m

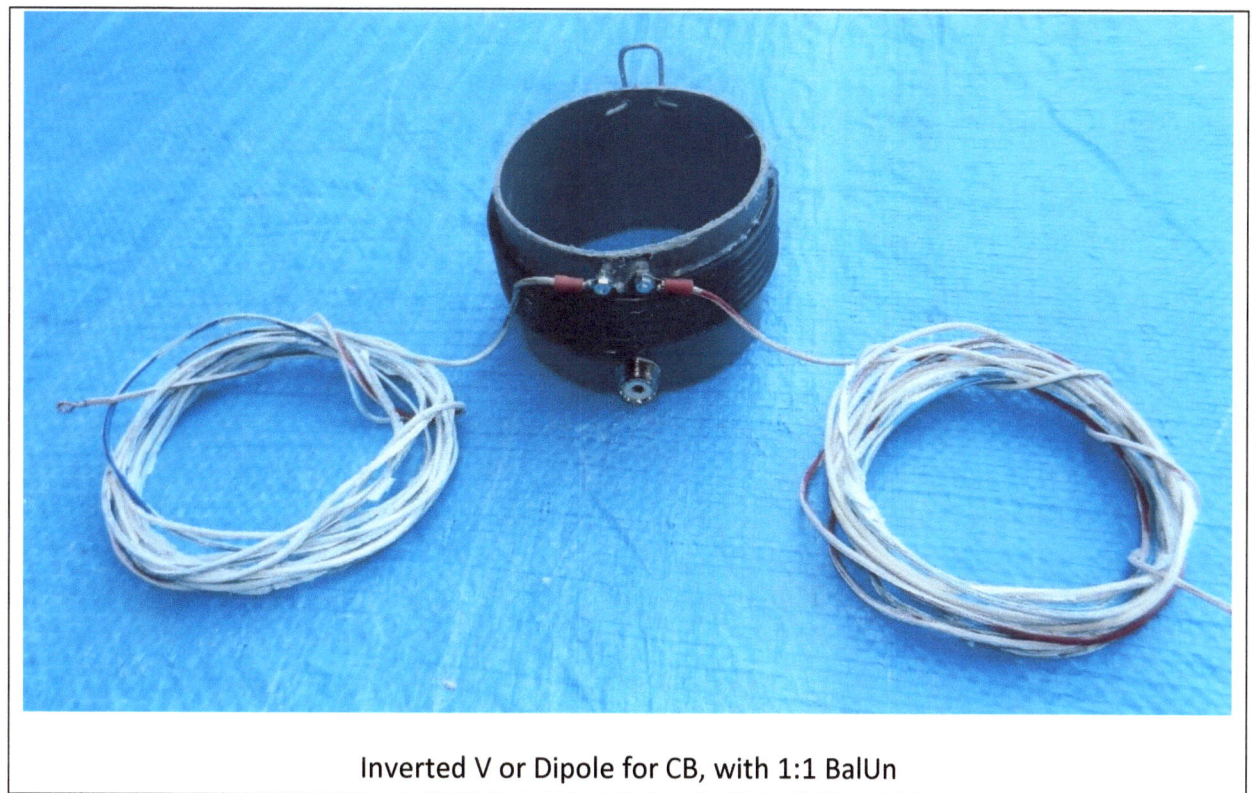

Inverted V or Dipole for CB, with 1:1 BalUn

10m (28.0-29.7 MHz)

Dipole, Inverted V and Delta Loop
For more information, see the Dipole section.

- Dipole total length = 5.09m, one arm length = 2.545m
- Inverted V total length = 4.84m, one arm length = 2.42m
- Delta Loop total length = 10.24m, one side length = 3.41m

Yagi 2
For more information, see the Yagi 2 section.

- Boom Length: 1.76m
- Dipole Length: 4.93m
- Reflector Length: 5.36m
- Impedance: 50 Ω

HB9CV
See the HB9CV section for all the data.

6m (50-54 MHz)

Dipole and Inverted V

For more information, see the Dipole section.

- Dipole total length = 2850mm, one arm length = 1425mm.
- Inverted V total length = 2710mm, one arm length = 1355mm

Ground Plane

For more information, see the GP section.

- Radial length = 147mm
- Vertical length = 140mm
- Radial material: aluminum pipe with diameter 8mm

HB9CV

See the HB9CV section for all the data.

2m (144-148 MHz)

Dipole and Inverted V

For more information, see the Dipole section.

- Dipole total length = 990mm, one arm length = 495mm
- Inverted V total length = 940mm, one arm length = 470mm

Ground Plane (GP)

For more information, see the GP section.

- Radial length = 52mm
- Vertical length = 49mm
- Radial material: aluminum pipe with diameter 6mm

J-pole and Slim Jim

Slim Jim is the improved variant of the J-pole antenna (short: JP, the other name is "Zepp"). It is omnidirectional half-wave antenna (half-wave is from the feeding point to the top, and quarter length from the feeding point to the bottom). Matching the impedance to the 50Ω and getting SWR close to 1:1 is obtained by sliding the connection of the feed line along the stub. It is essential for the antenna to be insulated and as far away as possible from the mounting structure. It has lower takeoff angle and better electrical performance than a 5/8 wavelength GP antenna. Its gain is 2.4dBi.

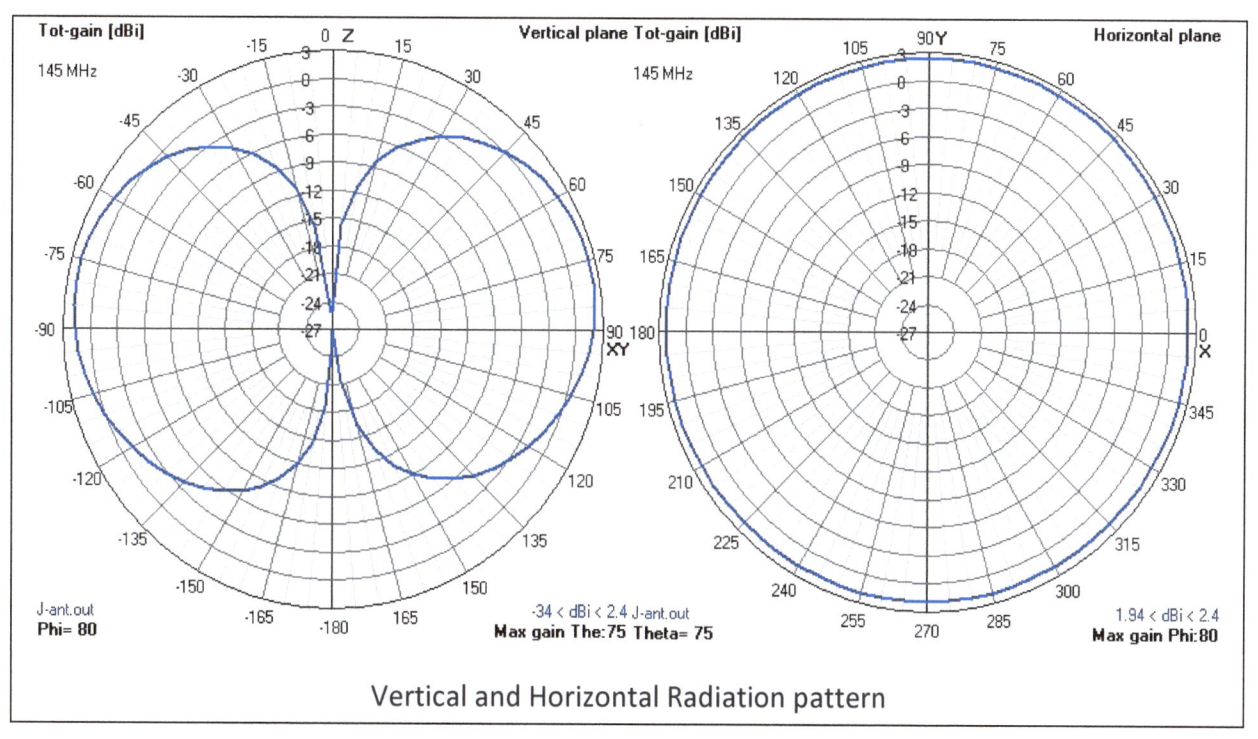

Vertical and Horizontal Radiation pattern

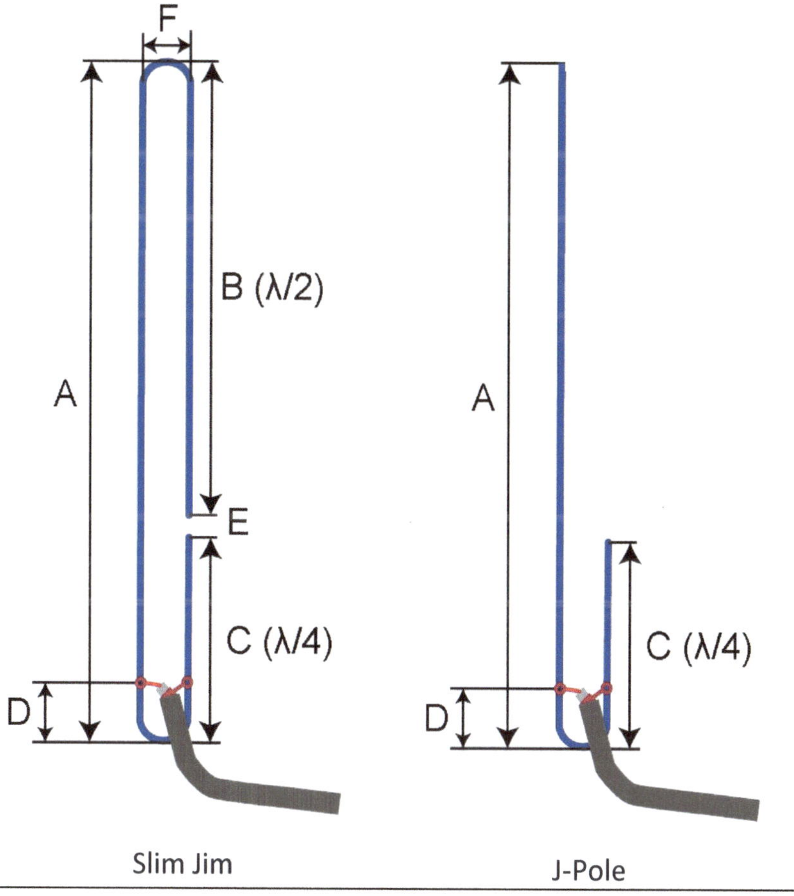

Measurements for the target frequency 145MHz and 50Ω impedance:

- A. Overall length – 1510mm (1490mm for J-Pole)
- B. Half-wave radiator section – 993mm
- C. Quarter wave matching section – 496mm
- D. 50Ω feed point – 103mm
- E. Gap – 20mm
- F. Spacing – 45mm

Slim Jim

J-Pole

D needs to be fine adjusted for the 50Ω. The given value is accurate in theory, however, since we cannot build a perfect antenna, the position for the 1:1 SWR will also be a bit away from ideal.

F is measured center from center of the pipe, and is not critical to be very accurate.

There is a need for 1:1 BalUn, for which the Ugly Choke can be used.

HB9CV

See the HB9CV section for all the data.

Yagi 7 with folded dipole

For the data and explanation refer to the Yagi section. Any number of elements can be used.

145MHz	Length	Spaced	Boom position	Insert to	Gain [dBi]	Boom
Reflector	1031	0	30	495.5	0	60
Dipole	999	414	444	497.5	4.3	474
1	946.2	155.1	598.6	453	6.9	628.6
2	937.4	372.2	970.7	448.5	8.6	1000.7
3	929.2	444.5	1415.2	444.5	9.9	1445.2
4	921.6	516.9	1932.1	441	11	1962.1
5	914.5	578.9	2511	437.5	11.9	2541
6	908	620.3	3131.3	434	12.7	3161.3
7	901.9	651.3	3782.6	431	13.3	3812.6
8	896.3	682.3	4464.9	428	13.9	4494.9
9	891.1	713.3	5178.2	425.5	14.4	5208.2
10	886.3	744.3	5922.5	423	14.9	5952.5
11	881.8	775.3	6697.8	421	15.3	6727.8

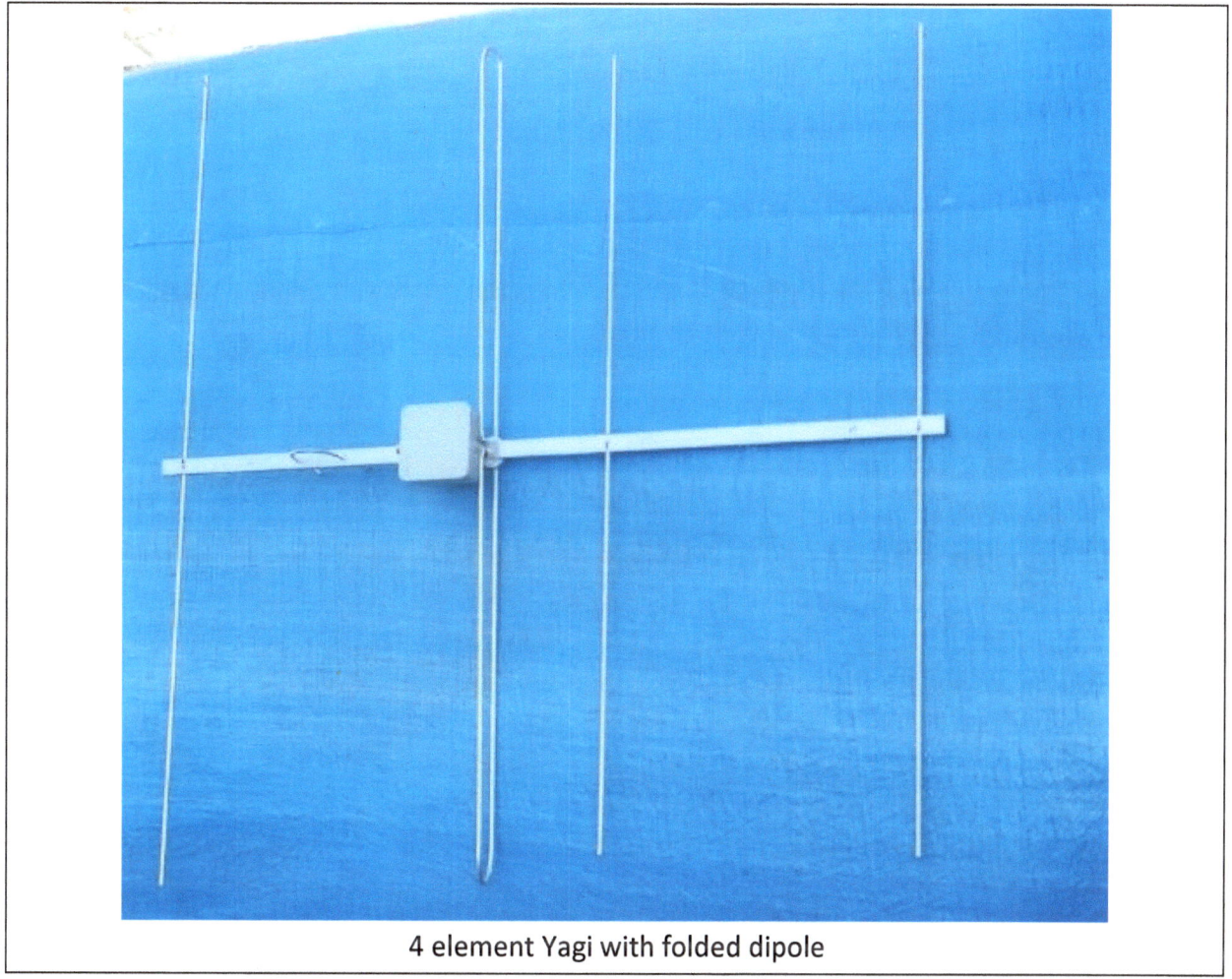

4 element Yagi with folded dipole

Yagi 6 28Ω

This has been tested to be a very good design. It is simple to make and is very reliable.

All elements should be made with Ø10 mm aluminum pipes.

The impedance of the antenna is 28Ω, so it is required to have BalUn with two 75Ω cables in parallel. See more about that in the BalUn section.

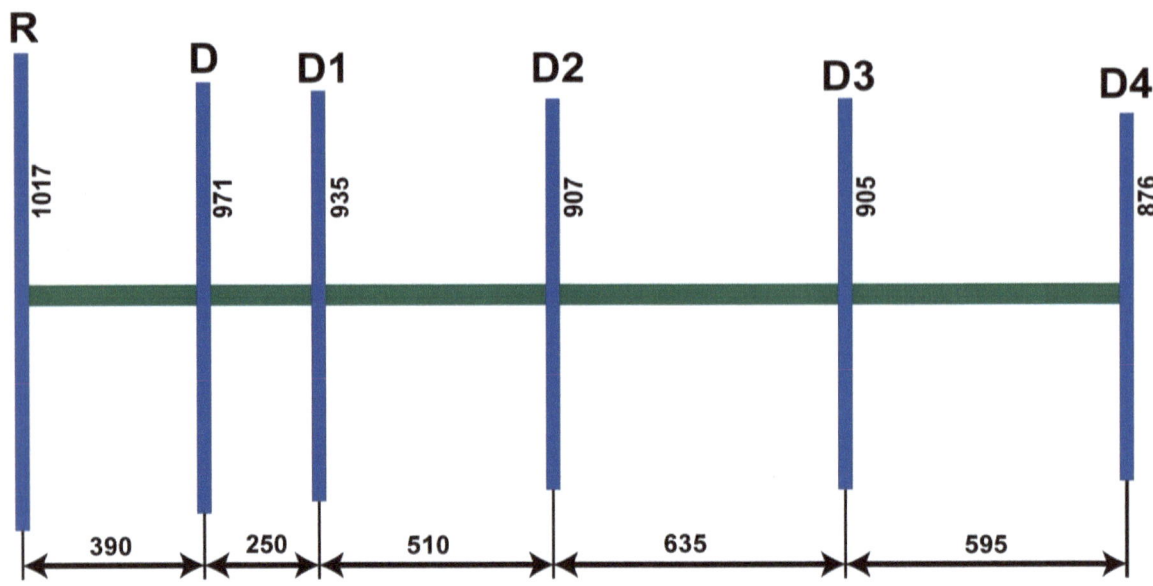

	Reflector	Dipole	D1	D2	D3	D4
Boom Position	0	390	640	1150	1785	2380
Spacing from previous	0	390	250	510	635	595
Length	1017	971	935	907	905	876

The critical for this model is that elements must be on the isolated standoff, but have connection with the metal boom on one central point. The screw should be Ø3 mm. Plastic isolator – standoff – should be maximally 3mm thick.

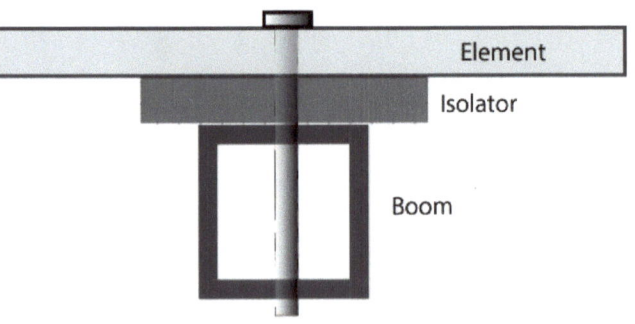

Co-Linear Omni – Franklin Antenna

The long name is "Linear Antenna Array with Series Excitation", but its popular name is Franklin Antenna. Gain of this antenna is 5.3dBi, and can be increased with adding additional A, B and C sections i.e. increasing the number of dipoles - A. Our antenna has 3 dipoles – two A sections and one D-D section.

Number of Dipoles	3	5	7	9
Gain	5.3	6.5	8.0	8.1

We can see that increasing the number of dipoles beyond 7 returns almost no increase in gain.

Its mechanism of work is very simple: it exploits the sinusoidal wave – when the positive part of the wave is in half-wave dipole A, the negative is in B-C-B section – the positive sections add up, while the negative radiation cancels itself – the gain is increased.

Antenna should be matched with 4:1 BalUn. It is good practice to place it in a PVC pipe.

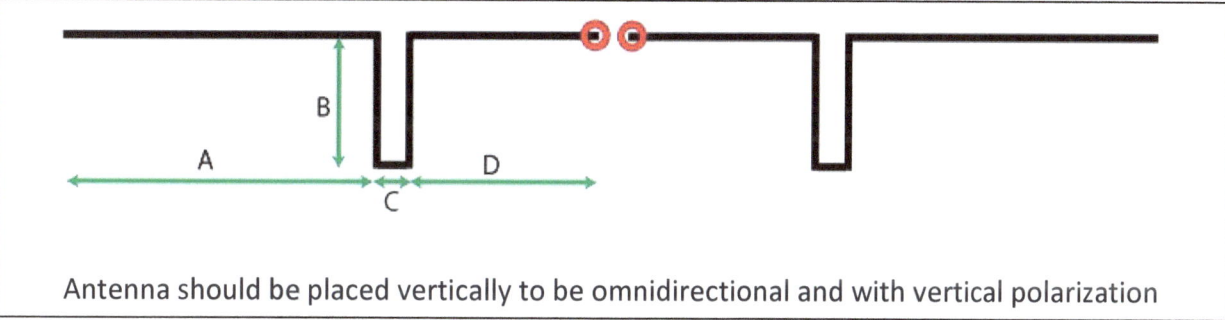

Antenna should be placed vertically to be omnidirectional and with vertical polarization

Measurements:

- A – half wave multiplied by velocity factor - 965mm,
- B – quarter wave - 508mm,
- C – 51mm or 10% from quarter wave,
- D – ½ A - 483mm.

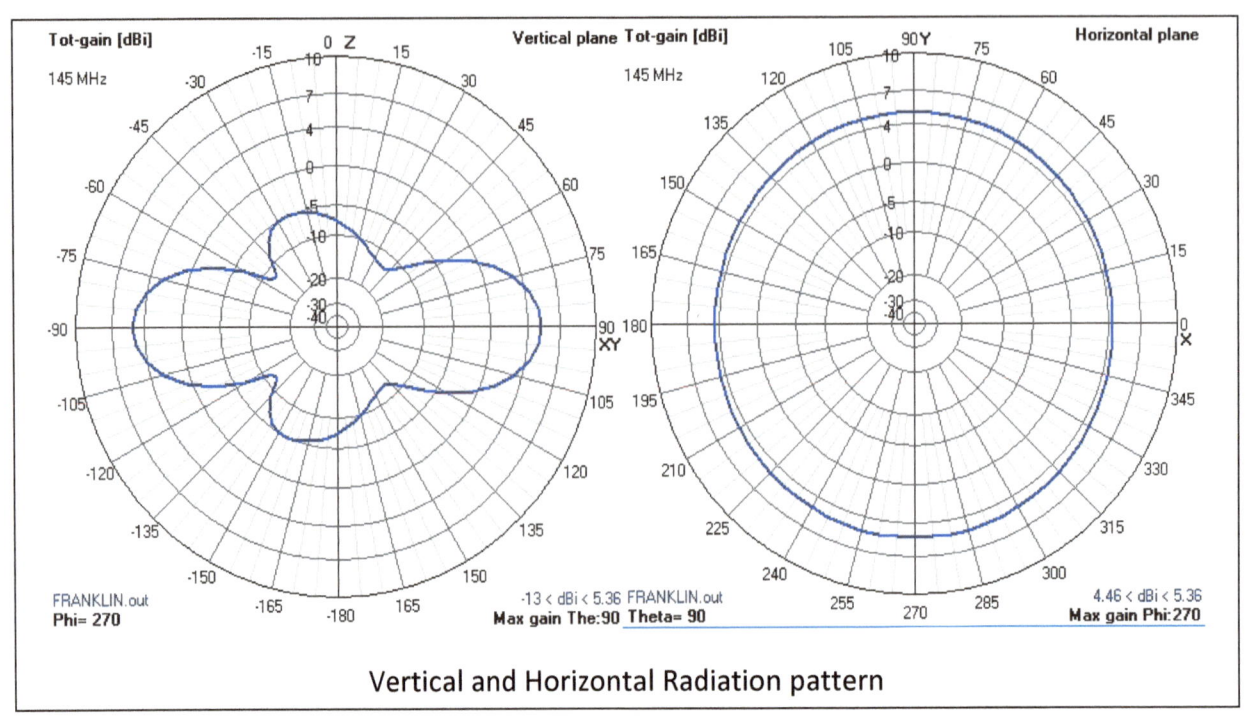

Vertical and Horizontal Radiation pattern

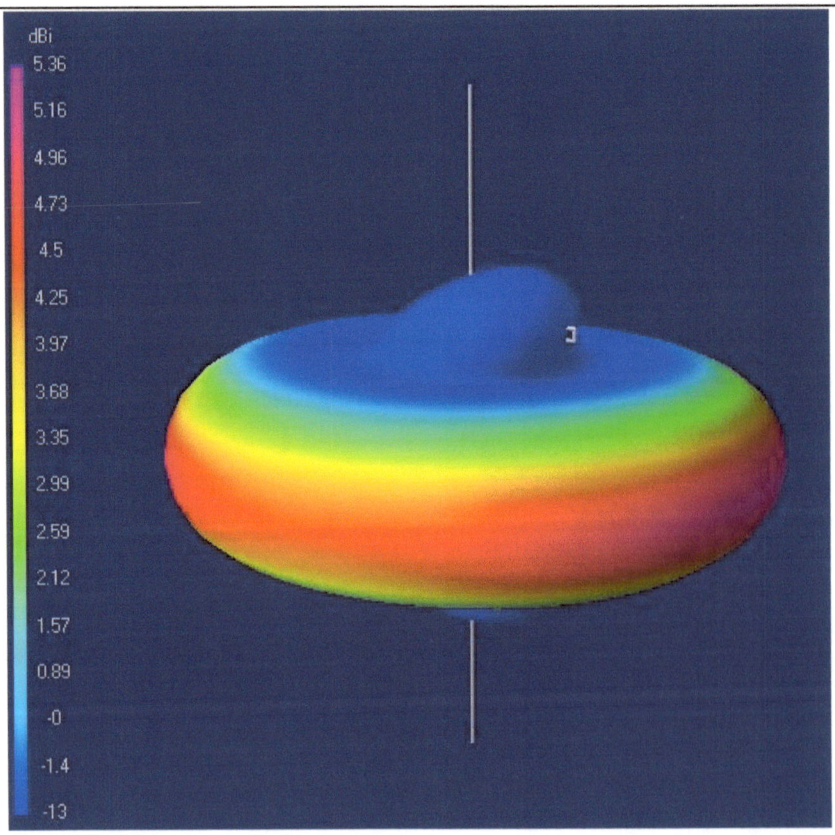

Radiation pattern in 3D

70cm (420-450 MHz)

Ground Plane (GP)

For more information, see the GP section.

- Radial length = 17mm
- Vertical length = 16mm
- Radial material: brass wire 3mm thick.

Yagi 15 with folded dipole

For data and explanation refer to the Yagi section. Any number of elements can be used.

435MHz	Length	Spaced	Boom position	Insert to	Gain [dBi]	Boom
Reflector	358	0	30	164	0	60
Dipole	149	138	168	149	4.3	198
1	318.4	51.7	219.5	144	6.9	249.5
2	314.8	124.1	343.6	142.5	8.6	373.6
3	311.4	148.2	491.7	140.5	9.9	521.7
4	308.3	172.3	664	139	11	694
5	305.5	193	857	137.5	11.9	887
6	302.8	206.8	1063.8	136.5	12.7	1093.8
7	300.4	217.1	1280.9	135	13.3	1310.9
8	298.1	227.4	1508.3	134	13.9	1538.3
9	296	237.8	1746.1	133	14.4	1776.1
10	294	248.1	1994.2	132	14.9	2024.2
11	292.2	258.4	2252.6	131	15.3	2282.6
12	290.5	265.3	2517.9	130.5	15.7	2547.9
13	289	268.8	2786.7	129.5	16	2816.7

Yagi 10 28Ω

Gain of this antenna with 10 elements is 15.2dBi. Since the impedance is 28Ω it must be matched with two 75Ω coaxial cables in parallel.

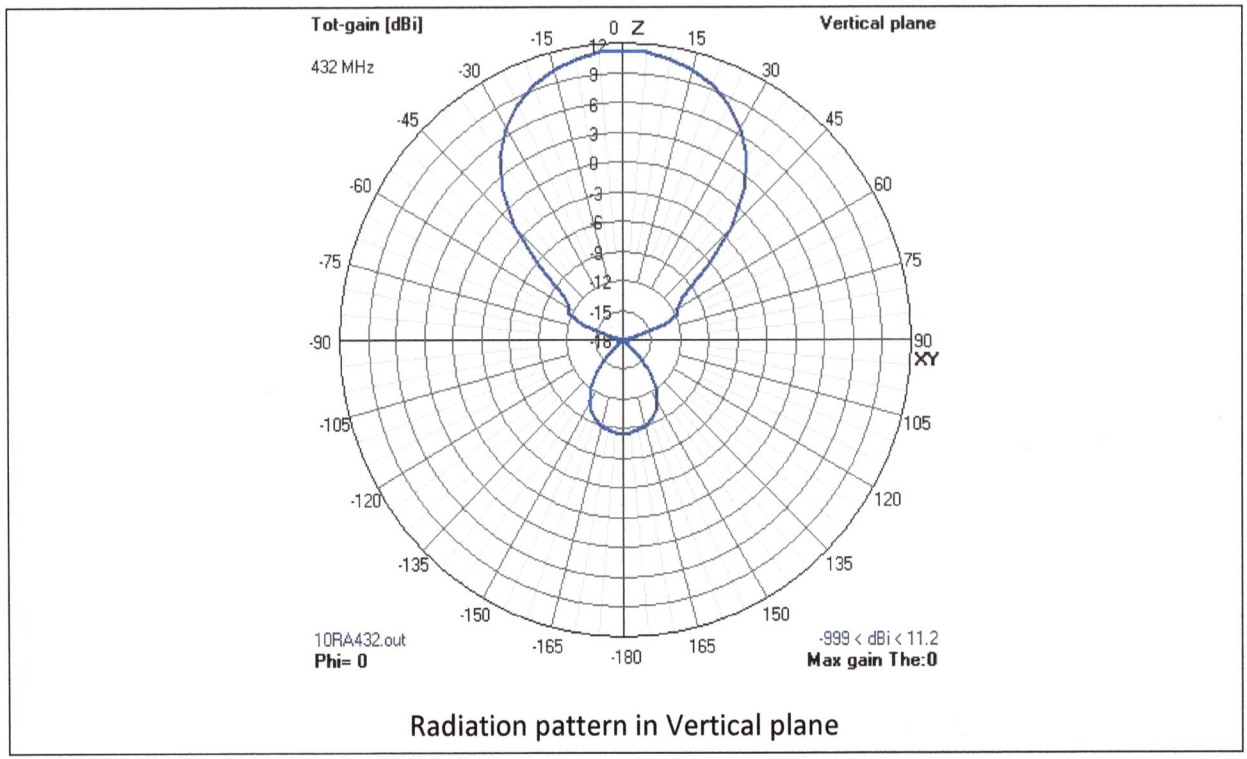

Radiation pattern in Vertical plane

All elements must be insulated from the Boom with standoffs.

The table shows the data with three pipe diameters: 10mm, 8mm and 6mm, but the dipole is in all three variants with Ø10 mm.[10]

Element	Element position [mm]	Ø10 mm [mm]	Ø8 mm [mm]	Ø6 mm [mm]
Reflector	0	327	330	331
Dipole	140	315	314 (Ø10 mm)	315 (Ø10 mm)
Director 1	235	300	302	306
Director 2	425	287	291	296
Director 3	650	281	285	290
Director 4	920	272	277	282
Director 5	1205	269	274	279
Director 6	1495	269	274	279
Director 7	1780	274	279	284
Director 8	1995	269	274	279

---

[10] Thicker pipe gives greater bandwidth of the antenna.

J-pole and Slim Jim

See the <u>Slim-Jim and J-pole for 2m band</u> for schematics and explanations.

Measurements for the target frequency 435MHz and 50Ω impedance:

A. Overall length – 503mm
B. Half-wave radiator section – 331mm
C. Quarter wave matching section – 166mm
D. 50Ω feed point – 17mm
E. Gap – 7mm
F. Spacing – 15mm

Biquad

For more information and schematics, see the <u>Biquad</u> section.

- Inner sides of the radiator should be 165mm long, outer sides 170mm.
- 1st choice for Reflector's width and height: 620mm × 620mm.
- 2nd choice: 1100mm × 827mm.
- Reflector should be made from net wire with density of at least 0.1λ. Because wind and ice can deform that type of reflector, some kind of frame is required.
- With the antenna BalUn 1:1 is desired.

Collinear Coax Omni

This antenna is very cheap to make and it requires no skill, only a lot of time to prepare all the coaxial sections of ½ wavelength.

If we choose RG-58U for our coaxial cable, we must account its velocity factor of 0.66. Thus, the ½ wavelength inside the cable will be 227mm.

We need to add 5mm on each side of the coax section – the part where there will be no shielding, only hot-wire – for the overall length of each section to 237mm. Braid shield should be exposed by at least 3mm on each side.

The last section should have hot-wire sticking out from it by ¼ wavelength, which is equal to 172mm. Thus we need to cut that section to 404mm, remove the braid shield from 232mm to the end.

At the bottom of the array we need Ø8mm copper tube that we slide over the coaxial cable – sleeve BalUn. Velocity factor for the tube is 0.95, thus the pipe for the ¼ wavelength should be 164mm long. The beginning of the tube should be welded to the shield.

To achieve 9dBi, we need to have 8 sections.

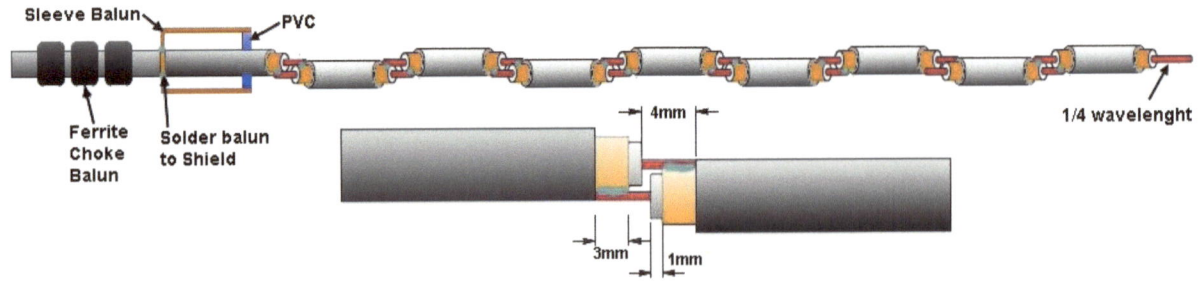

1:1 BalUn is desired. It can be achieved by placing toroid ferrite cores over the coaxial cable ½ wavelength below the feed point.

After each section is welded, open parts should be hot-glued. After that, the whole antenna should be placed in a PVC pipe.

23cm (1240-1300 MHz)

Super Yagi 24

This antenna differs a bit from the one described in the Yagi section. It has 24 elements and 18.2dBi gain. The model described before is with directors bonded through metal boom. Here we will present the design that uses standoffs. This is the recommended design if standoffs are available.

The target frequency is 1296MHz. It should cover the whole 23cm band.

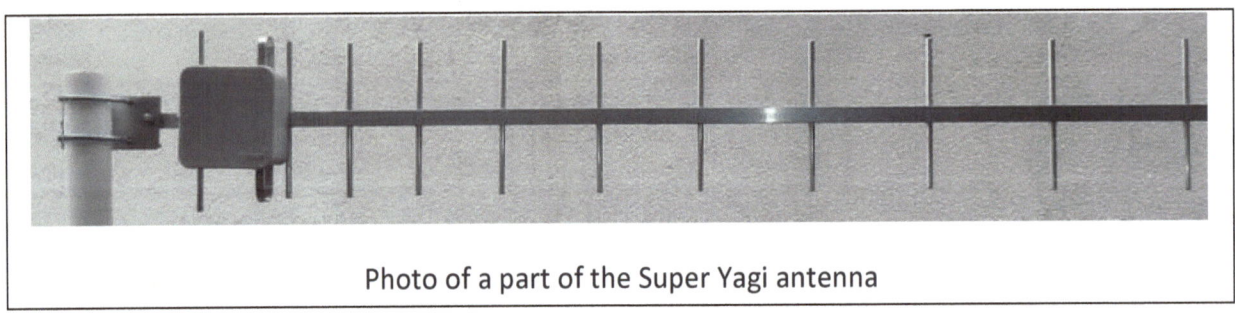

Photo of a part of the Super Yagi antenna

Direct or	Length [mm]	Spaced [mm]	Boom Position [mm]	Insert to [mm]	Gain [dBd]	Gain [dBi]
1	93.1	17.3	93.6	39.5	4.8	6.9
2	91.6	41.6	135.3	39.0	6.5	8.6
3	90.2	49.7	185.0	38.0	7.8	9.9
4	89.0	57.8	242.8	37.5	8.9	11.0
5	87.8	64.8	307.6	37.0	9.8	11.9
6	86.7	69.4	377.0	36.5	10.5	12.7
7	85.7	72.9	449.8	36.0	11.2	13.3
8	84.8	76.3	526.2	35.5	11.7	13.9
9	83.9	79.8	606.0	35.0	12.2	14.4
10	83.1	83.3	689.3	34.5	12.7	14.9
11	82.4	86.7	776.0	34.0	13.1	15.3
12	81.7	89.1	865.1	34.0	13.5	15.7
13	81.0	90.2	955.3	33.5	13.8	16.0
14	80.4	91.4	1046.7	33.0	14.2	16.3
15	79.9	92.5	1139.2	33.0	14.5	16.6
16	79.4	92.5	1231.7	32.5	14.7	16.9
17	78.9	92.5	1324.2	32.5	15.0	17.1
18	78.5	92.5	1416.8	32.0	15.2	17.4
19	78.1	92.5	1509.3	32.0	15.4	17.6
20	77.7	92.5	1601.8	32.0	15.6	17.8
21	77.3	92.5	1694.4	31.5	15.8	18.0
22	77.0	92.5	1786.9	31.5	16.0	18.2

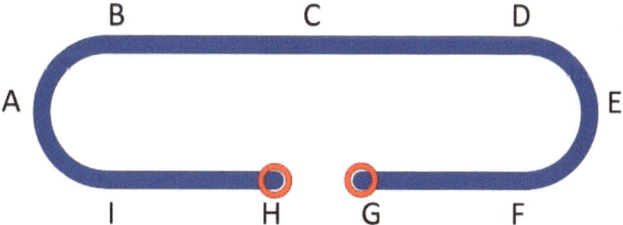

Other measurements:

- Diameter of dipole bend (outer diameter) of the ribbon used when constructed – BI and DF - 10mm. Dipole gap at feed point - the distance H-G – 5mm. Dipole should be made out of flat ribbon: width – 10mm, thickness – 1mm. Measured tip-to-tip, i.e. from A to E, it should be 107mm long. It should be spaced from reflector by 46mm, which is at boom position of 76mm. Total ribbon's length – 232mm, BC = CD = 44mm, HI = GF = 41mm, HA = GE = 57mm, HB = GD = 73mm, HC = GC = 116mm.

- Boom should be with square cross section with one side 14mm, 1817mm long. There are overhangs of 30mm before the reflector and after the last director. Antenna should be mounted at the middle of boom.
- Directors should be made of Ø6 mm aluminum pipes.
- BalUn should be 4:1, made from RG-58 coaxial cable.

Biquad

For more information and schematics, see the <u>Biquad</u> section.

- Inner sides of the radiator should be 54mm long, outer sides 58mm.
- 1st choice for Reflector's width and height: 240mm × 240mm.
- 2nd choice: 370mm × 277mm.
- Reflector should be made from net wire with density of at least 0.1λ or aluminum plate. Because wind and ice can deform that type of reflector, some kind of frame might be necessary.
- With the antenna BalUn 1:1 is desired.

Wi-Fi

2.4GHz

Patch

Patch is very simple directional antenna used mostly for higher frequencies. It consists of one metal patch – radiating element (that replaces and acts like a dipole), and reflector. Gain can be as high as 9dBi – the patch itself acts like a two dipoles (gain of cca. 5-6dBi), and the reflector adds additional 3dBi. Its simplicity and low cost are the reasons why many manufacturers are producing it for the Wi-Fi.

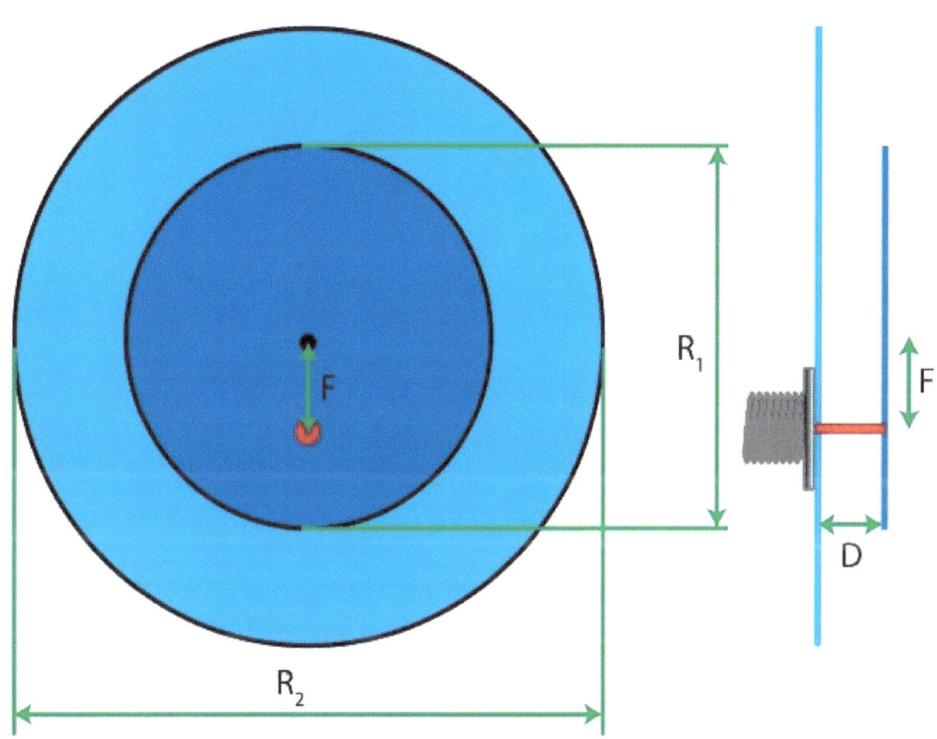

- R_1 – diameter of the radiating element (patch),
- R_2 – diameter of the reflector,
- F – distance between the feeding point and the center of the radiating element/reflector,
- D – distance between radiating patch and reflector.

Cable Impedance	R_1 [mm]	R_2 [mm]	F [mm]	D [mm]
50Ω	65	100	17	7
75Ω	65	100	20	9

Although for all applications the 50Ω cable is recommended, its cost might defer from using it for the antenna that is meant for the short distances. 75Ω RG-6 cable might be enough for such applications. In that case some of the parameters must be altered: *F* and *D*, to make a proper impedance match.

Center of the connector must be welded to the patch, while the mass of the connector should be screwed onto the reflector.

The holders for the patch should be insulators, all but the one in the center, which can be a metallic screw. Around it at least three of the insulating holders must be applied to secure the patch.

Cantenna

Cantenna is a popular name for a waveguide[11] antenna built from a tin can that can ideally have 8.5dBi gain.

Advantages of the antenna:

- Materials are widely available,
- Easy to make,
- Small,
- Can be used as feed-point for parabolic antenna,
- Cheap, only cost of the antenna is for the connector.

Disadvantages:

- Gain is significantly smaller than for Biquad antenna,
- As a feed for parabolic antennas, Biquad is by far superior,
- It must be enclosed on the front with an insulator, so that water cannot enter the can,
- Mounting the antenna can be a challenge.

The can diameter is ideally ½ wavelength. In practice, for Wi-Fi 2.4GHz (2.412GHz – 2.484GHz), a tin cans with diameter from 3" (76mm) to 3.5" (89mm) is fine, ideally 3.25" (82mm). Tin can with diameter in that range will cover the whole band.

The length of the waveguide can should be at least ¾ wavelength, which is equal to 190mm.

---

[11] Waveguide is a transmission line for EM waves, built from metal of specific size and shape. The inner surface of the waveguide directs the signal.

Radiator must be placed at the exact distance from the back of the can. Front of the can is open so that the wave can enter the can, back is closed so that it can reflect it back to the monopole.

Monopole should be long ¼ wavelength (63mm) and should be placed at the same distance from the back of the tin can. The wire used should be brass wire with 4mm diameter. It must be welded to the center of the connector. Any type of female connector can be used: N connector is the best choice, but it is expensive, F connector is cheap but has loses at high frequencies, BNC is much better than F and is 4x less price than N, so it is a good compromise.

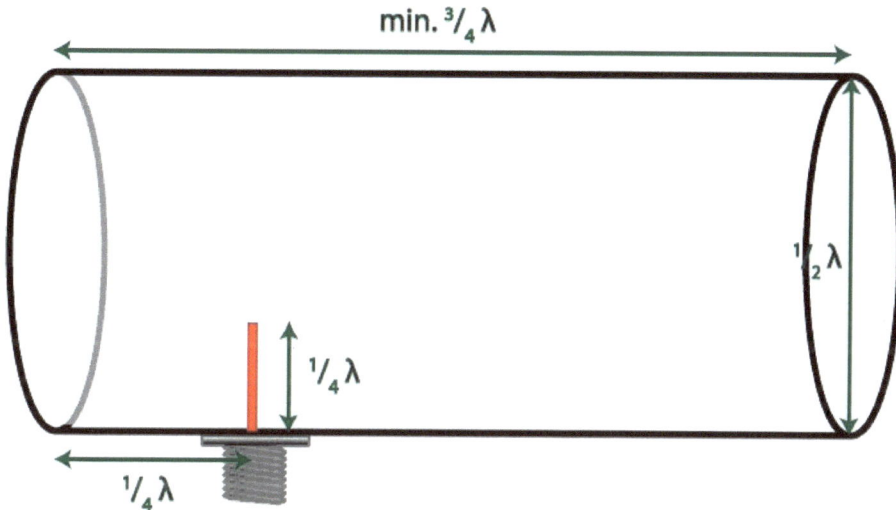

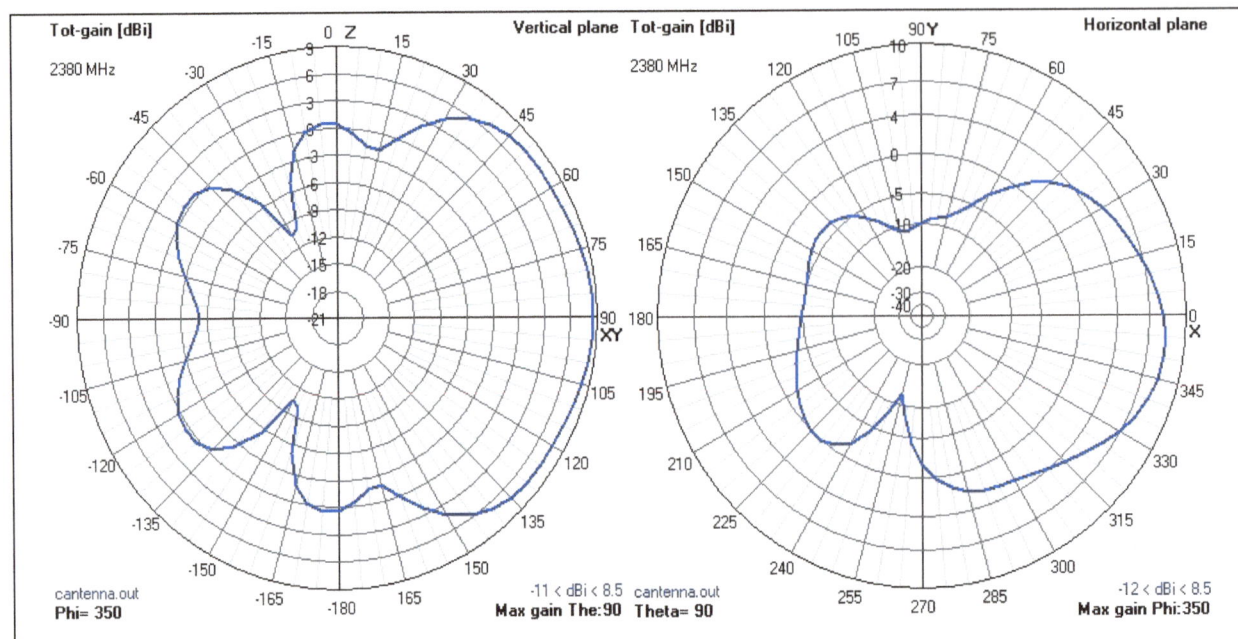

Radiation pattern in Vertical and Horizontal plane; image below – in 3D

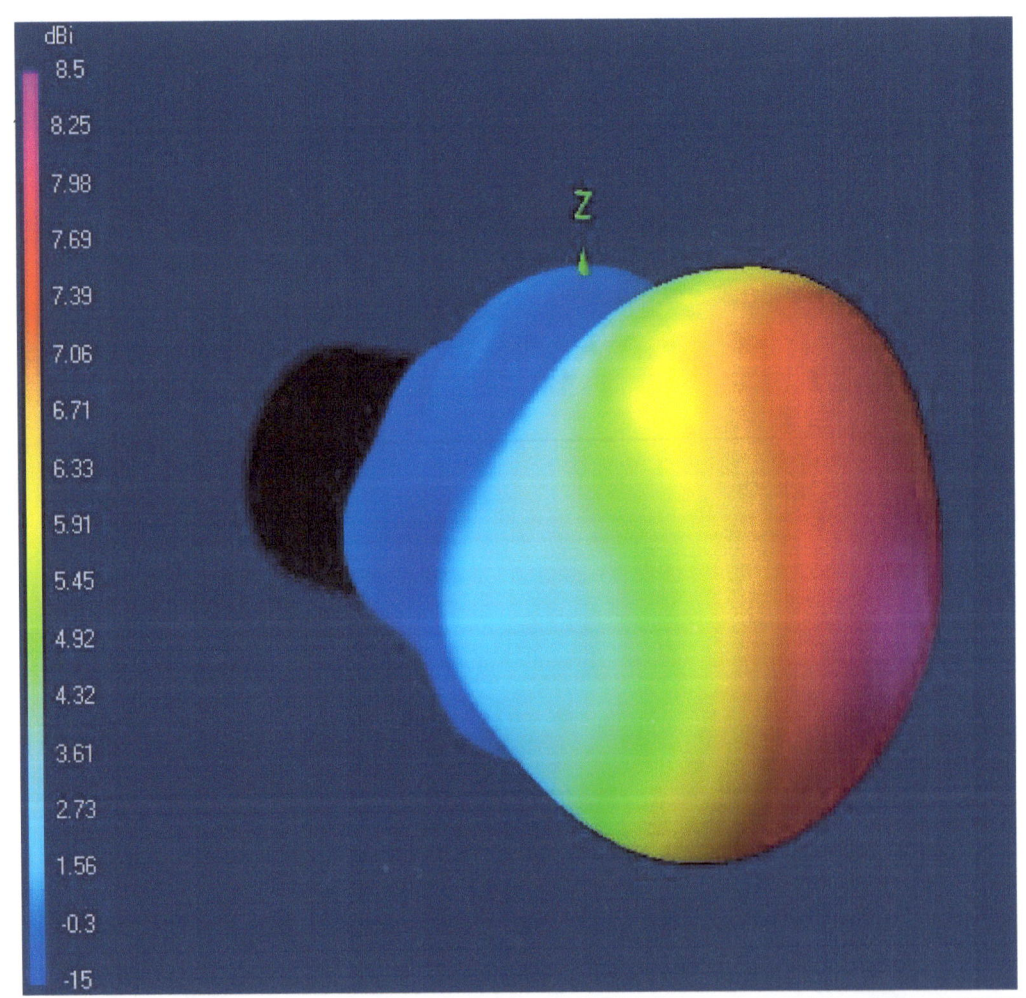

Biquad

For more information and schematics, see the <u>Biquad</u> section.

- Total wire length should be 241mm,
- Inner sides of the radiator should be 29mm long, outer sides 31mm,
- 1st choice for Reflector's width and height: 110mm × 110mm,
- 2nd choice: 200mm × 150mm,
- Reflector should be made from aluminum plate, at least 0.5mm thick.

Helix

Helical antennas are widely used on almost all frequencies, however, because it is hard and expensive to make proper mounts for the helix at low frequencies, we will describe only one helical antenna – for 2.4GHz. It differs from the other antennas described mainly in its polarization – it is circular – which has its drawbacks and advantages:

- Circular polarization gets better around the objects that interfere with the signal, especially trees,
- Circular polarization is rarely used, so two circularly polarized antennas beamed one at another will have low interference from the others, making it suitable for crowded areas,
- In works with both vertically and horizontally polarized antennas, but with 3dB loss in signal,
- It does not work if the windings of the antennas are opposite, i.e. if one helix goes clockwise and the other counterclockwise (right-hand and left-hand circular polarizations) – their circular polarizations will also go into opposite directions and the end result would be 20dB loss,
- Gain is small comparing to Yagi, for the same length,
- Building one such antenna is cheap,
- Proper mounting the spring to the reflector is hard (birds, snow and gravity are its enemies).

Theory

The antenna is a coil with *N* turns and a reflector – ground. The circumference *C* of one turn should be equal to one wavelength and the distance between the turns should be ¼ of circumference. Reflector should be a circle or a square, with diameter (for circle) or side (for square) equal to one wavelength.

The gain can be obtained from this equation (Kraus approximate formula; all units given in meters):

$$G = 11.8 + 10 \log\left(\left(\frac{C}{\lambda}\right)^2 10Nd\right)$$

For a diameter of 40mm, and wire thickness 2mm, circumference will be:

$$\pi \cdot (40 + 2)\text{mm} = 132\text{mm}$$

which is equal to λ, ($d = 33$mm) so the gain equation can now be simplified to:

$$G = 11.8 + 10 \log(10Nd)$$

Let's say we want antenna to have 12 turns. According to the equation, the gain would be 17.7dBi. However, the equation is inaccurate, as we can see from the 4Nec2 simulation below, the gain is 3dB less than predicted. It should be 14.7dBi. Thus the equation should be corrected to:

$$G = 8.8 + 10 \log(10Nd)$$

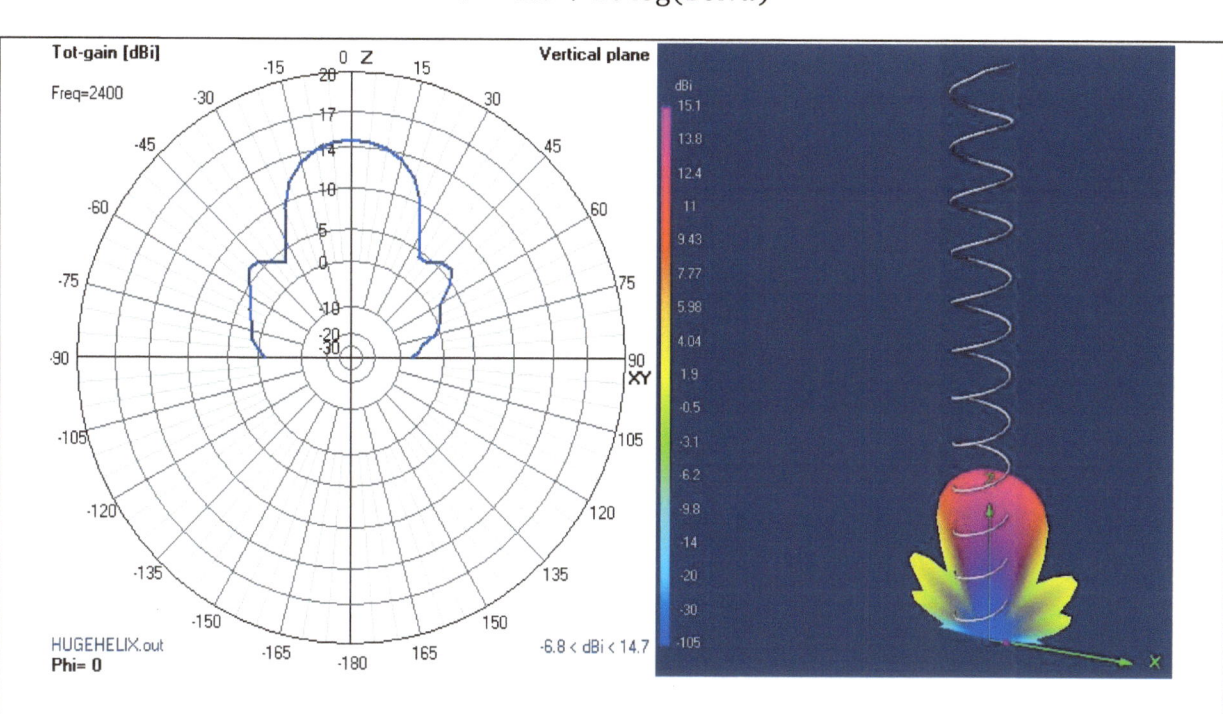

Radiation pattern of Helix antenna with 12 turns

Impedance Matching

The impedance of the antenna is approximately 150Ω, thus a proper matching to 50Ω is required. This can be easily done with 71mm × 17mm right-angle triangle made from copper foil or plate.

Practical design

Materials needed:

- Copper foil: 71mm × 17mm right triangle,
- PVC pipe with Ø40mm and 40cm long,
- Aluminum square with side 130mm or circle with diameter 130mm,
- PVC end cap for PCV pipe Ø40mm outer diameter,
- Copper wire $2mm^2$, insulated, best with varnish insulation – enameled wire,
- Some insulator over the whole pipe is recommended, but not necessary.

Procedure:

1. Cut the pipe to 40cm,
2. Drill a small Ø2mm hole at the end of the pipe,

3. Mark a straight line through that hole, along the pipe,
4. Mark on it a points on every 33mm,
5. Mark a straight line on the opposite side,
6. Mark on it points on every 33mm, but starting from 16mm from the beginning,
7. Put the tip of the wire through the hole, and start winding the coil up to the beginning of the pipe. Wire should pass over the marked lines on both sides. As you move along, make sure to hot-glue the wire to the pipe, or it might unwind by itself,
8. Cut the wire at the beginning of the pipe,
9. Cut the excess wire that got through the hole at the end of the pipe,

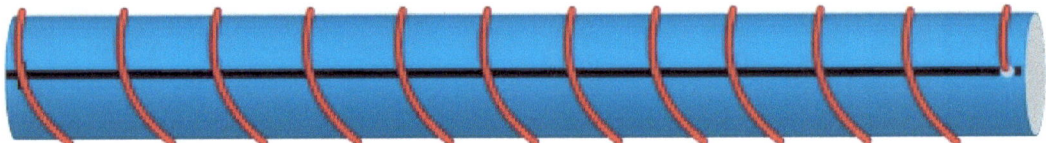

10. Put the endcap on the end of the pipe,
11. Drill other endcap through the center and mount it firmly to the aluminum plate's center with a screw and hot-glue,
12. Drill it on one side just next to reflector, with Ø10mm drill bit (this is a hole where we will weld copper foil with connector,
13. Mount connector through reflector on the place where that hole is,

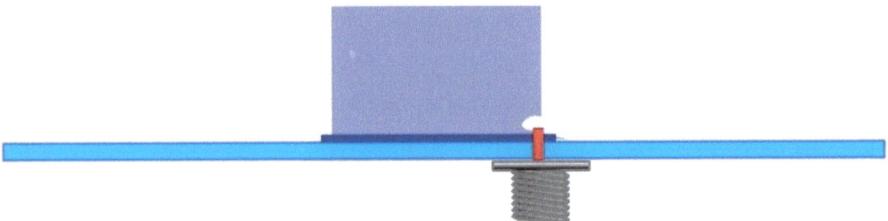

14. Cut a right-angle triangle with sides 71mm and 17mm,
15. From the beginning of a pipe and coil, place the triangle so that the shorter side goes along the line on the pipe, so that the longer side ends on the coil,
16. At that place cut the wire and remove insulation,
17. Weld the wire to the end of the longer side of triangle,
18. Place the pipe in the PVC cap and secure it with hot-glue gun,

19. Weld triangle to the connector,

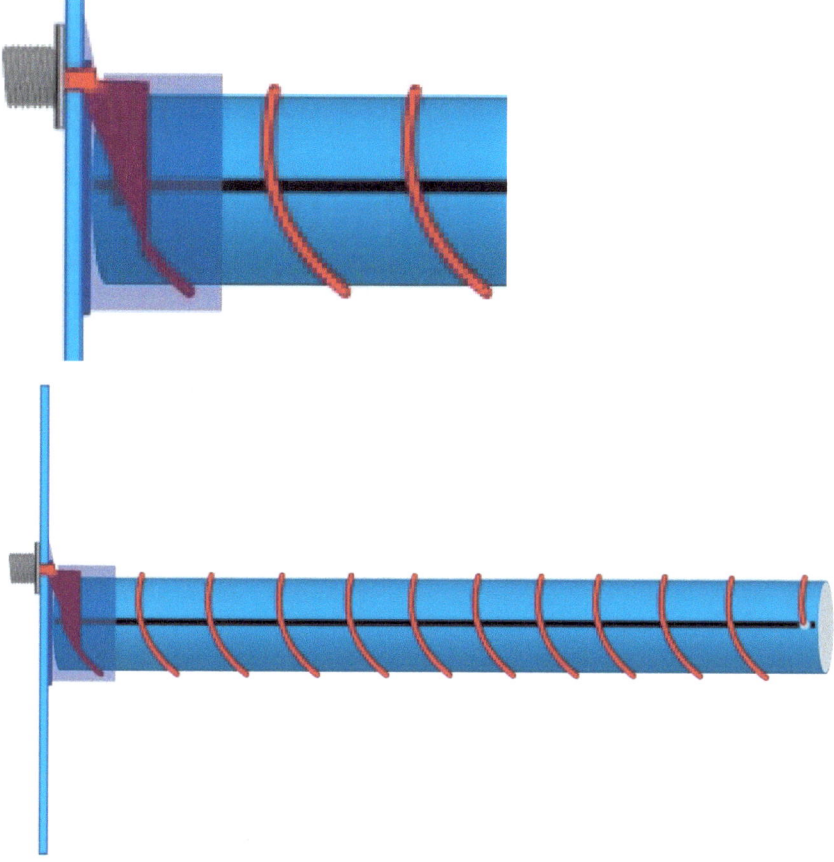

20. To improve isolation, it is the best to use Ø50 mm heat-shrinks (but they are expensive and hard to find), or at least paint or PVC tape.

3D corner
50Ω

More about 3D corner can be found in the 3D corner section.

To the place where director should be mounted, it is necessary to drill the reflector and use the brass screw. We then weld our director, shortened for the height of the screw's head, to the screw. This way there is a good electrical contact between reflector and director.

The thickness of the director should be 4mm.

Connector's mass must be in good electrical contact with the reflector as well.

To insulate antenna, hot-glue gun the connector and lower part of the director, and paint them with non-metallic paint (metallic paints have aluminum particles in them that can affect the antenna's performance).

75Ω

To save on material, we can make a compromise and use four times cheaper RG-6 cable (than RG-58). The director is not needed in this case, nor is connector, as we can strip the braid shield from the coaxial cable at the length of a monopole, and put it into PVC housing, like on the photo below. The hot wire of the cable becomes monopole, and shield should be connected to the reflector.

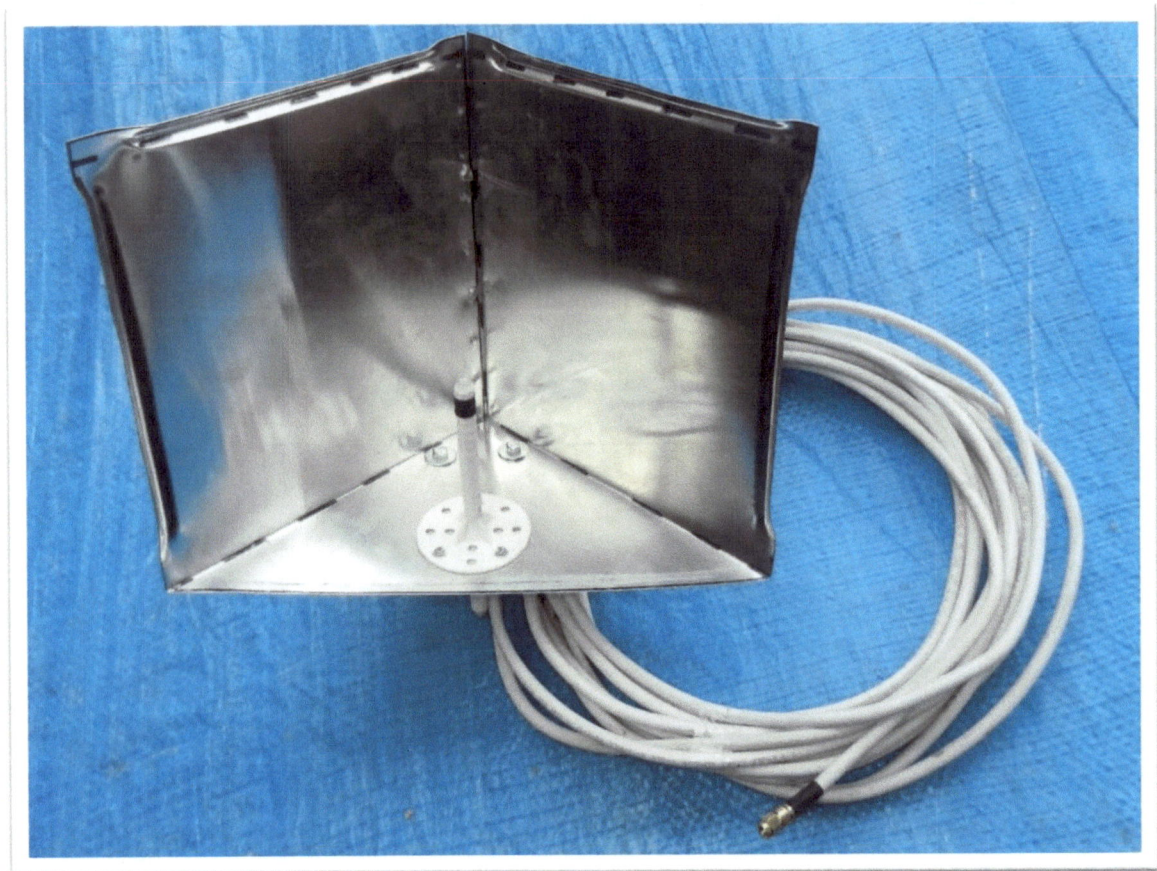

Sector V

The credit for this antenna goes to *Dragoslav Dobricic*, who made modifications of the Franklin dipole to be used as a sector antenna in the Wi-Fi 2.4GHz band, and named the antenna *Amos*.

Sector antennas got their name from the radiation pattern – narrow vertical and wide in horizontal plane. They are good for internet providers, in fact, it is better to use several Sector antennas than one omni directional. The gain of that type of antennas usually comes from the narrowing of the vertical radiation. Maximum gain is 12.5dBi.

This antenna is based on the Franklin collinear dipole, described in the 2m section.

Impedance of the antenna is 200Ω, thus it is ideal to use <u>4:1 BalUn</u>.

Reflector should be made from aluminum plate, 1mm thick, and wide ½ wavelength. Wider reflector would result in narrowing the horizontal radiation pattern and the antenna would become more directional.

Every dimension is in mm and measured from the center of the conductor. The distance between the reflector and upper part of the Franklin dipole is 29.5mm, i.e. 10mm from the lower parts.

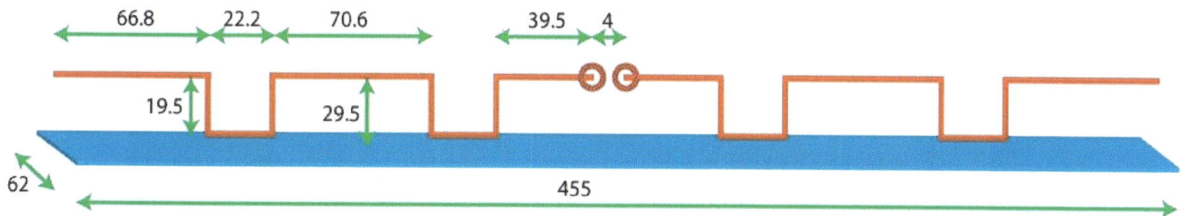

The antenna should be placed vertically to radiate vertical polarization and for the wide horizontal pattern.

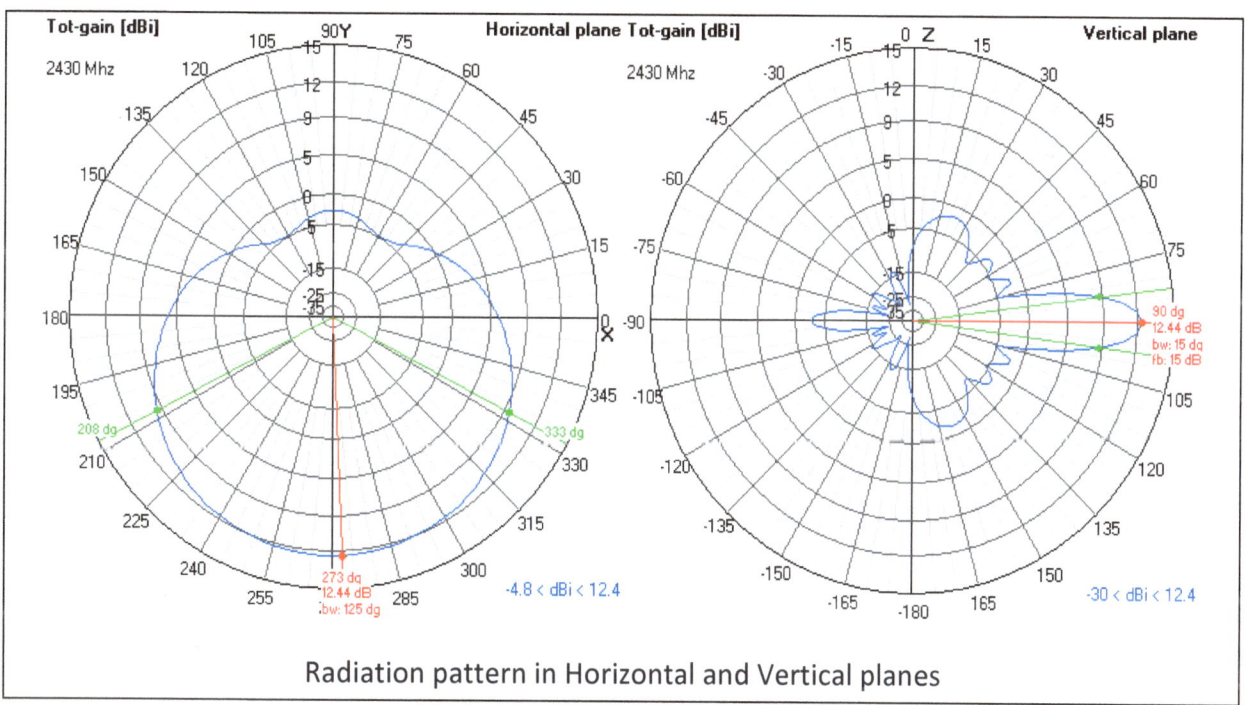

Radiation pattern in Horizontal and Vertical planes

Sector H

The credit for this antenna goes to *Dragoslav Dobricic*, who made modifications of the Biquad antenna to be used as a sector antenna in the Wi-Fi 2.4GHz band, and named the antenna *Quados*.

The antenna has 16.5dBi gain and horizontal polarization.

Wire should be of Ø2mm and total length is 1310mm. This exceeds the 1m of brass wire sold in hardware stores, thus we must make it from two 655mm segments that need to be welded later on. Reflector must be made from aluminum plate at least 0.5mm, dimensions 73mm × 540mm. Impedance is 200Ω thus we need 4:1 BalUn to match it to 50Ω.

Standoffs for the radiating wire must be from an insulator. Drill a hole just enough for a cable or connector at the center of the reflector.

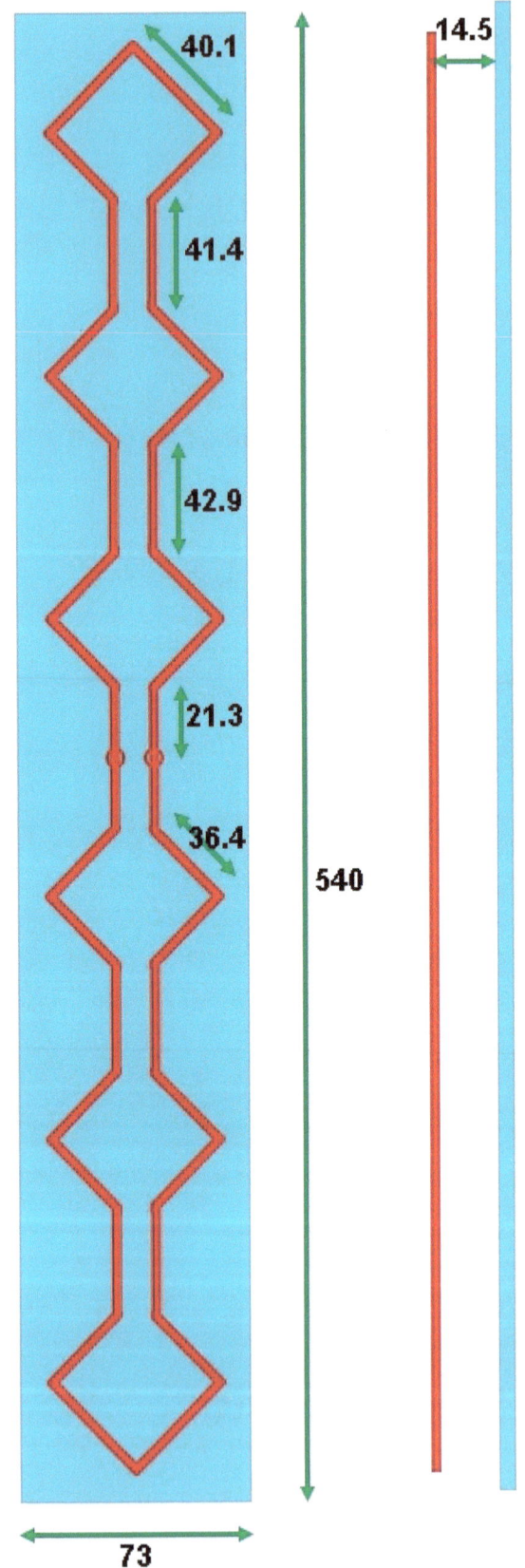

Panel PCB

Gain of the antenna is very good 14dBi for its size. It is commercially used by many manufacturers. It consists of the four patches in front of a reflector and a matching circuit. The impedance is 50Ω.

The image on the next page shows the print for the PCB (it is up to scale). The distance between the reflector and the copper side of PCB should be 7mm. Standoffs should be plastic; metallic standoffs will result in a drop in performance. The connector's hot-lead should be welded to the center of the vertical copper lead on the PCB (image is oriented for horizontal polarization, vertical can be achieved by simply rotating antenna by 90°.

Reflector should be made from aluminum plate 1mm thick, square shape 200mm × 200mm. Mass of the connector should be screwed through the hole on the center.

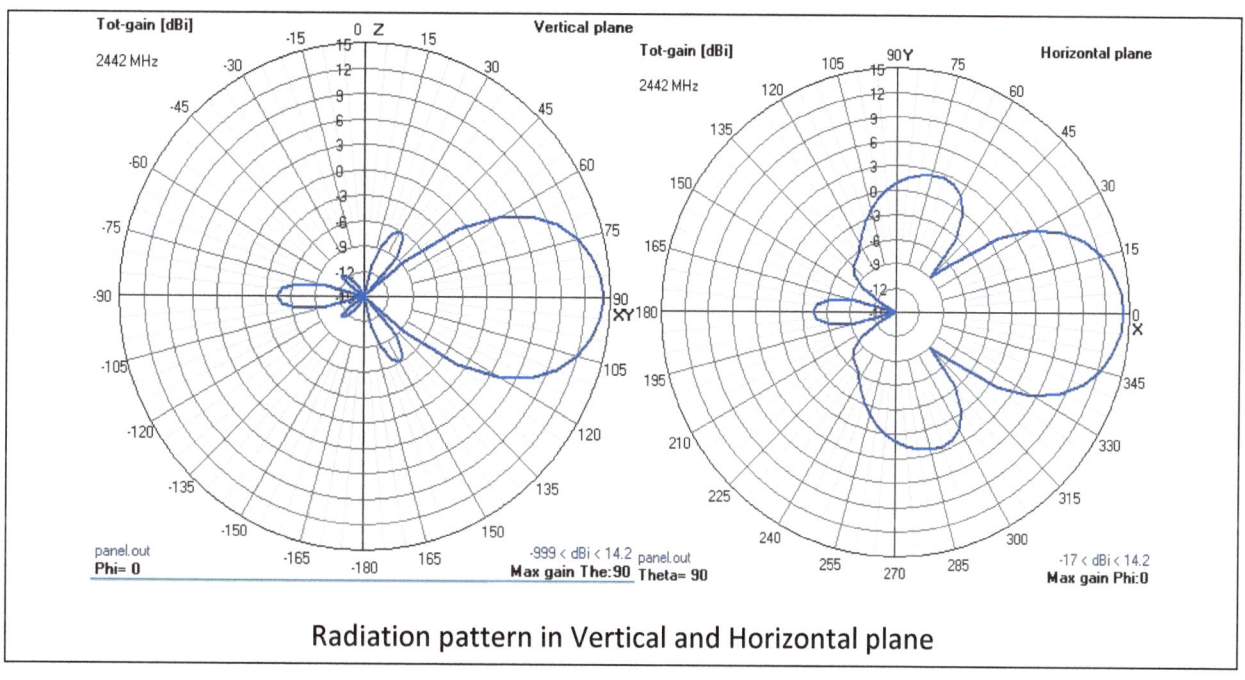

Radiation pattern in Vertical and Horizontal plane

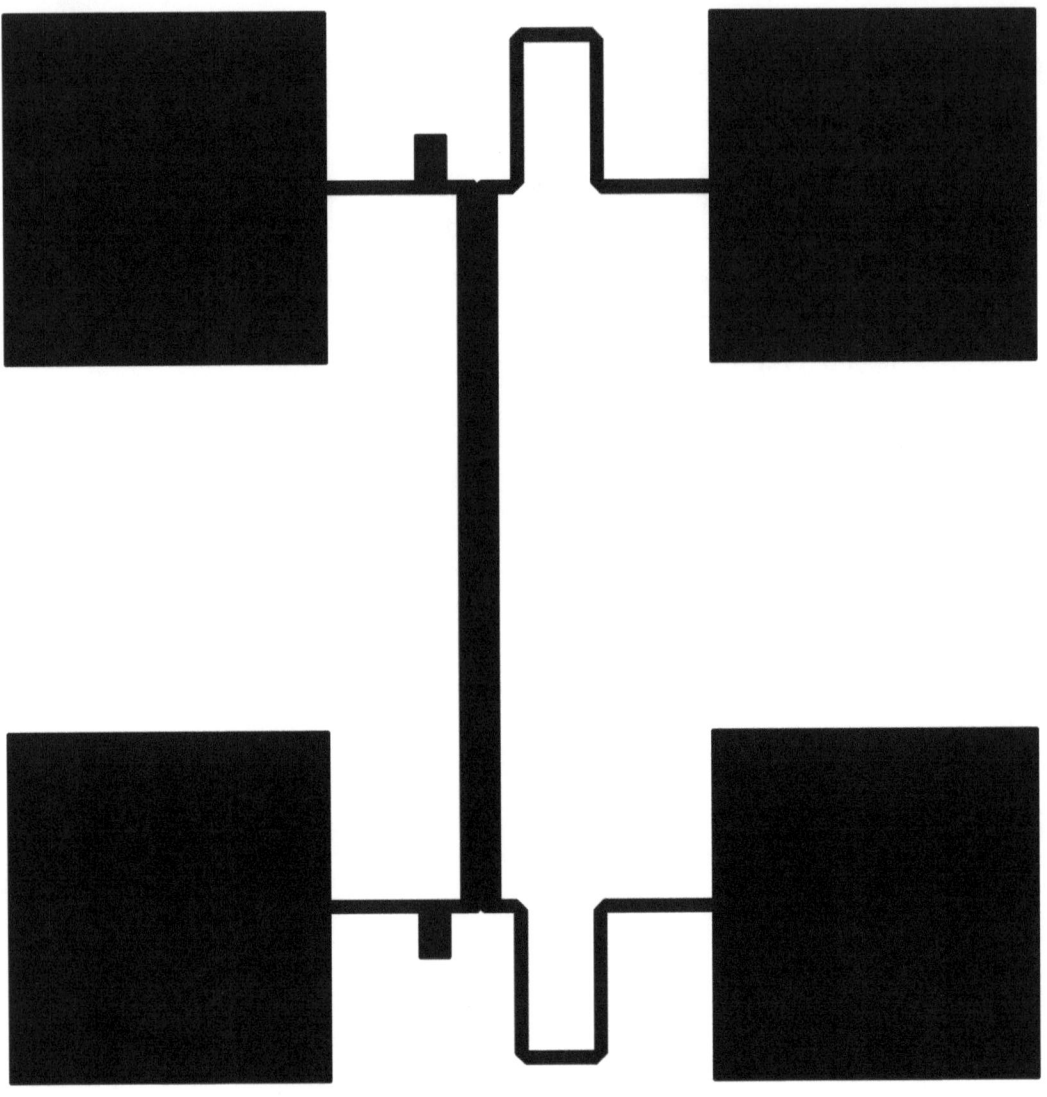

Print for transfering to PCB

32 slotted Waveguide Omni and 16 slotted Waveguide Sector

The credit for the design goes to *Trevor Marshall*.

This antenna is complicated and expensive to build, but the results it gives are very much worth the trouble. Its gain is 15dBi if built to exact measures, and the polarization is horizontal only. If rotated for vertical polarization, antenna would cease to be omni directional. It has almost equal horizontal radiation pattern in all directions, and very narrow vertical pattern.

At the bottom of the antenna there is a monopole. That monopole radiates signal through our waveguide reflector. Signal escapes where the slots are, and those slots then act like dipole antennas. It is obvious that slots must be cut on both sides of the antenna for it to be omnidirectional. Slots cut on only one side will make it very good sector antenna with 17dBi, however, Quados is much cheaper and easier to produce, with the same results. The antenna has no match if made as omni, other types of omni antennas cannot give such results.

Reflector's square pipe must have 3mm thick walls. Dashed lines represent endplates, placed inside the waveguide.

If we choose other size of the reflector than on the image, with dimensions 100mm × 50mm × 3mm instead of 4"× 2" × 1/8", the feed point must be offset by 15mm from the center line instead of 10mm in order for SWR to be closer to 1.

On the back, the slots should be on the opposite sides (it looks like the front side reflected horizontally), i.e. one can see through the antenna through the slots.

The monopole is made from a copper plate. Rolled up and welded, it turns into a cone, whose tip is to be welded to the hot lead of the N connector.

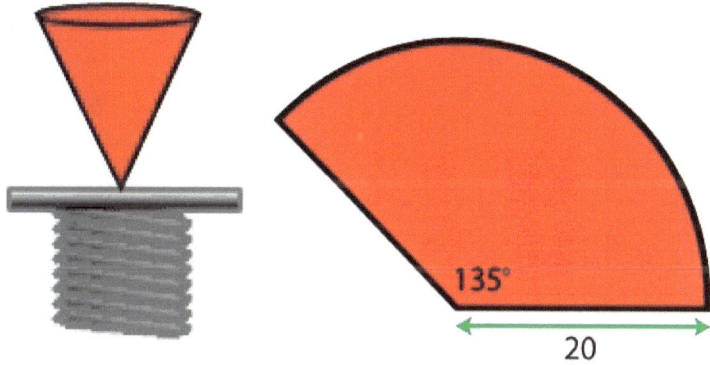

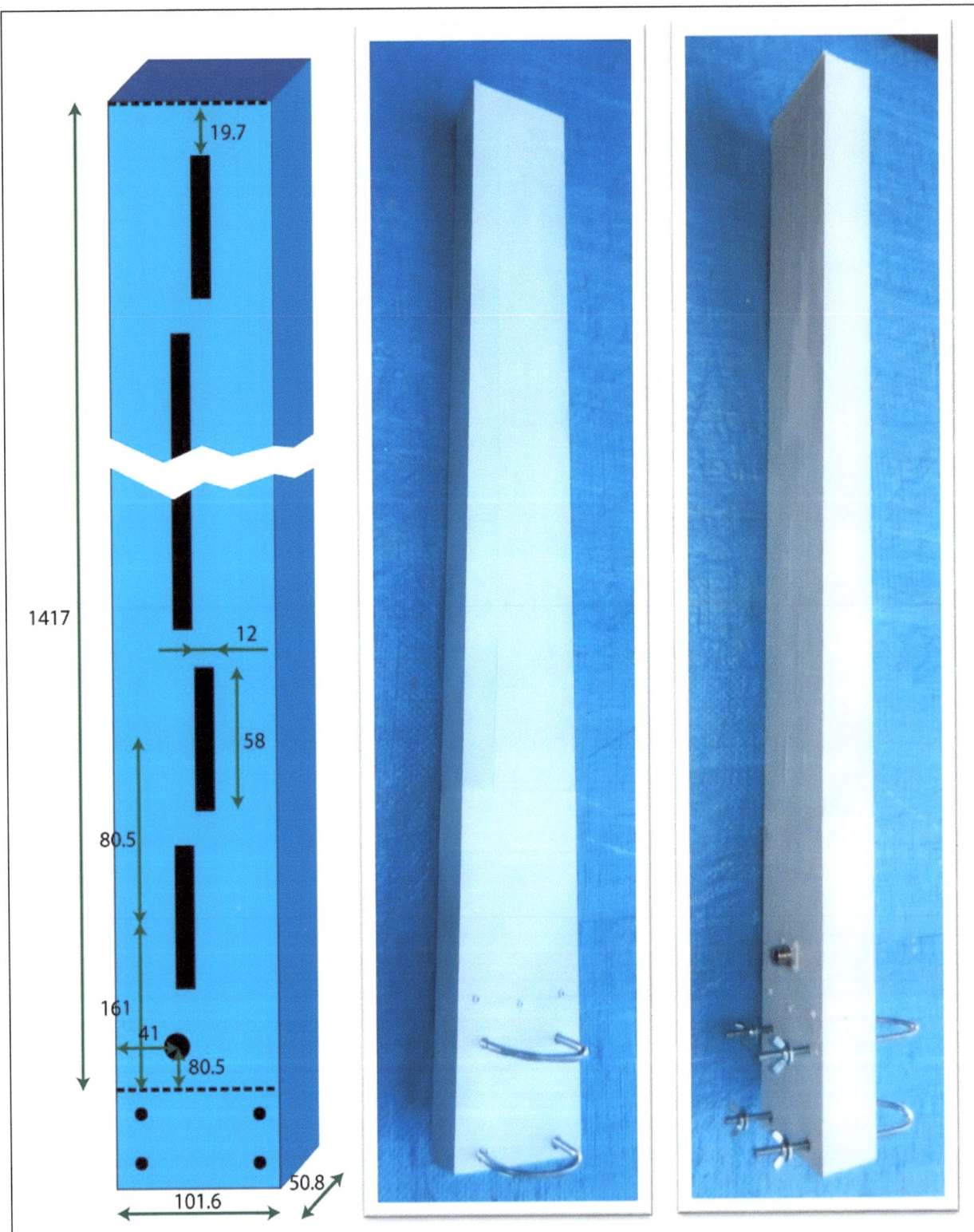

Left: Blueprints for the antenna (not up to scale); Middle and Right: finished product with U-bolts for mounting to the mast and covered slots

5GHz

Biquad

For more information and schematics, see the <u>Biquad</u> section.

- Total wire length should be 110mm,
- Inner sides of the radiator should be 12.5mm long, outer sides 15mm,
- 1st choice for Reflector's width and height: $51\text{mm} \times 51\text{mm}$,
- 2nd choice: $93\text{mm} \times 68\text{mm}$,
- Reflector should be made from aluminum, copper or brass plate, at least 0.5mm thick.

Sector Pillbox 14dBi

Pillbox antenna has multiple methods of work: vertical parallel plates are used as a waveguide, back of the reflector is parabolic and the feed is monopole. The distance between the vertical parallel plates should be just less than ½ wavelength. ¼ wavelength monopole must be placed in the focus of the parabolic reflector.

The antenna has very wide horizontal radiation pattern (-3db lines are wide 100°) and narrow vertical pattern, with 14dBi gain.

Reflector should be made from either brass or copper[12] plate thick at least 0.5mm. Image below should be transferred to the plate and cut to exact measures. After two side plates are made, there are also back and front reflector plates that need to be welded all the way to the side plates.

Connector must be welded all the way around to the reflector.

Monopole is made from copper or brass wire, thick 2mm and long 11.5mm (measured from the reflector surface).

Whole antenna should be placed inside PVC case, if used outdoors.

---

[12] Brass is recommended because copper is too soft, and aluminum is out of the question because it cannot be welded with tin.

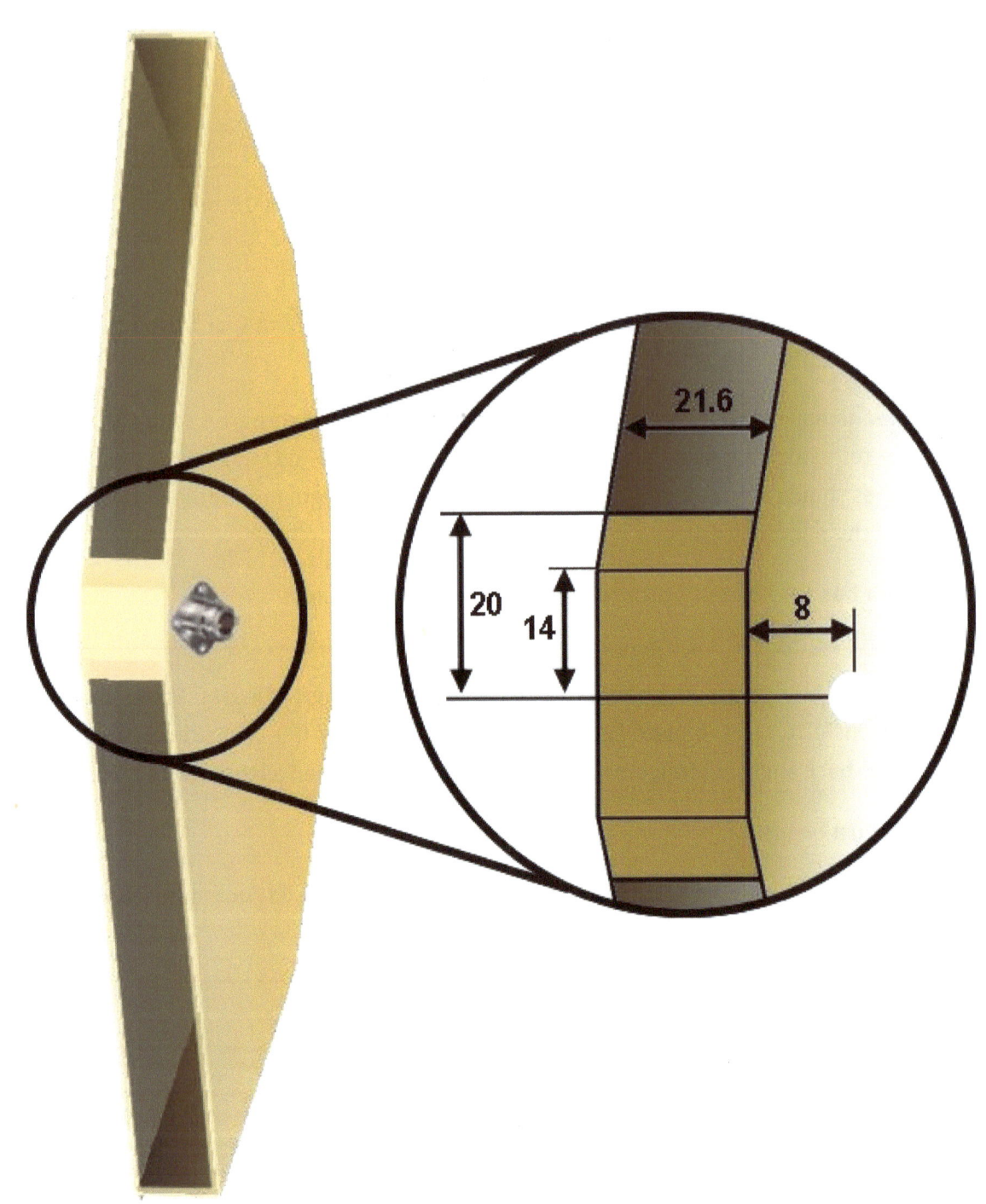

Print to transfer to brass plate, the images are up to scale

3D Corner

More about 3D corner can be found in the <u>3D corner</u> section.

- Reflector side should be 162mm for 18dBi,
- Reflector's aluminum plate thickness: 0.5mm,
- Monopole length: 41mm,
- Director length: 35mm,
- Monopole distance from reflector surfaces: 32mm,
- Director distance from reflector surfaces: 38mm,
- Brass wire thickness: 2mm.

GSM/GPRS

More about Yagi's can be seen in the Yagi section.

Yagi 12 PVC Boom

The element dimensions for directors and reflector should be shorter for the PVC boom than for the Yagi with metallic boom and elements bonded through (Yagi presented below), for the metallic boom width.

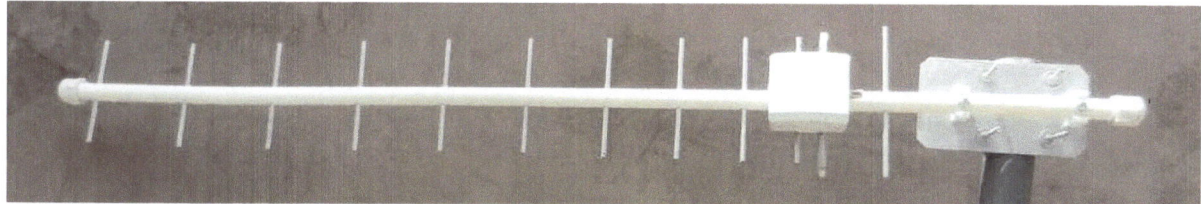

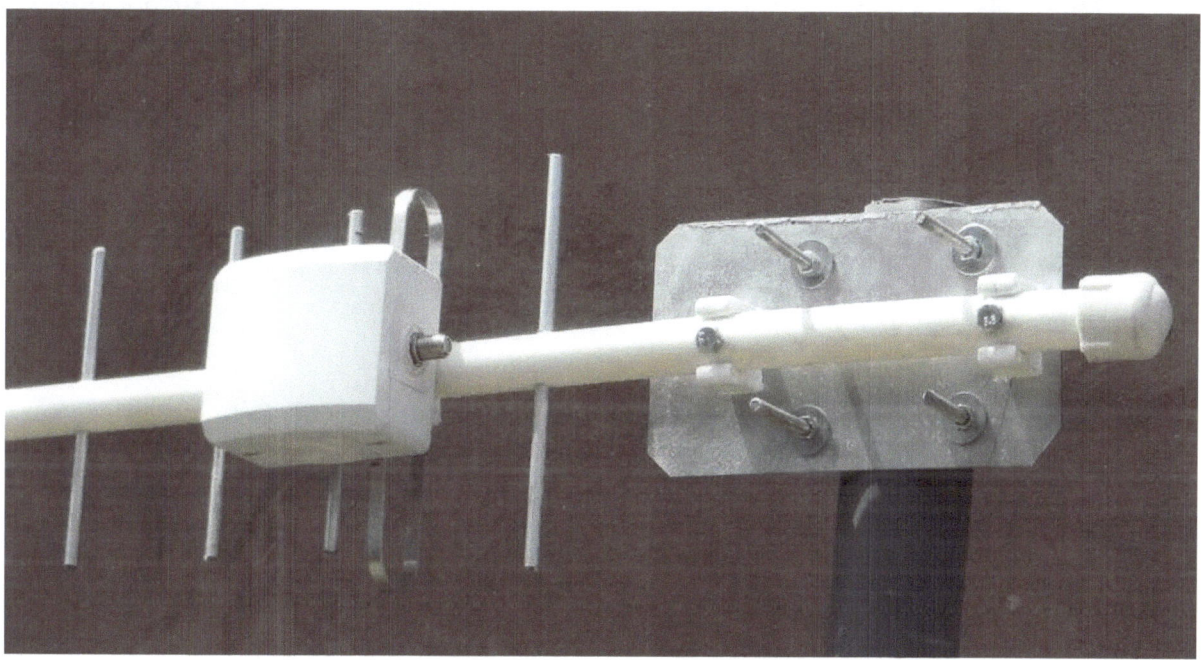

Yagi 12 metallic boom

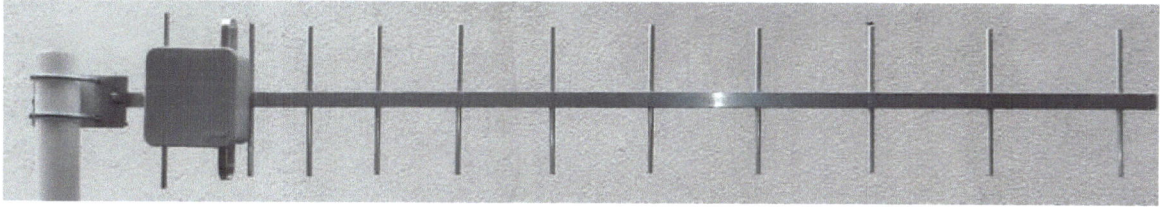

850MHz

850MHz	Length	Spaced	Boom position	Insert to	Gain [dBi]	Boom
Reflector	185.9	0	30	83	0	60
Dipole	165.7	71	101	71	4.3	131
1	161.3	26.5	127	70.5	6.9	157
2	159.2	63.5	190.5	69.5	8.6	220.5
3	157.3	75.8	266.3	68.5	9.9	296.3
4	155.5	88.2	354.5	68	11	384.5
5	153.9	98.8	453.2	67	11.9	483.2
6	152.4	105.8	559	66	12.7	589
7	150.9	111.1	670.1	65.5	13.3	700.1
8	149.6	116.4	786.5	65	13.9	816.5
9	148.4	121.7	908.2	64	14.4	938.2
10	147.3	127	1035.2	63.5	14.9	1065.2
11	146.2	132.3	1167.4	63	15.3	1197.4
12	145.2	135.8	1303.2	62.5	15.7	1333.2
13	144.3	137.6	1440.8	62	16	1470.8

920MHz

920MHz	Length	Spaced	Boom position	Insert to	Gain [dBi]	Boom
Reflector	172.8	0	30	76.5	0	60
Dipole	152.7	65	95	66.5	4.3	125
1	149.6	24.4	119.6	65	6.9	149.6
2	147.7	58.7	178.3	64	8.6	208.3
3	145.9	70.1	248.3	63	9.9	278.3
4	144.2	81.5	329.8	62	11	359.8
5	142.7	91.2	421	61.5	11.9	451
6	141.2	97.8	518.8	60.5	12.7	548.8
7	139.9	102.6	621.4	60	13.3	651.4
8	138.6	107.5	729	59.5	13.9	759
9	137.5	112.4	841.4	59	14.4	871.4
10	136.4	117.3	958.7	58	14.9	988.7
11	135.5	122.2	1080.9	57.5	15.3	1110.9
12	134.6	125.5	1206.4	57.5	15.7	1236.4
13	133.7	127.1	1333.4	57	16	1363.4
14	132.9	128.7	1462.2	56.5	16.3	1492.2
15	132.2	130.3	1592.5	56	16.6	1622.5

BalUn

Intro

BalUn converts between impedances and symmetry of feed-lines and antennas. In simple terms, we use BalUn when impedance of the cable and antenna or device's output are different. When we fed balanced antenna with coaxial cable, which is unbalanced, we also need to use a BalUn. BalUn prevents cable from becoming part of the antenna and radiating power.

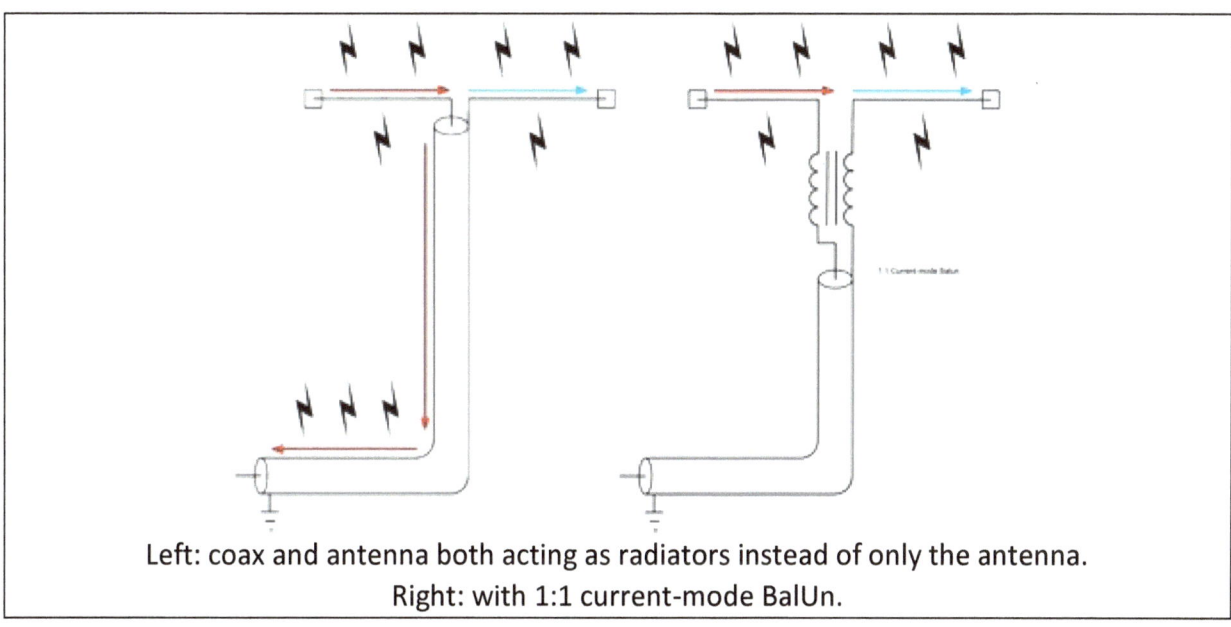

Left: coax and antenna both acting as radiators instead of only the antenna.
Right: with 1:1 current-mode BalUn.

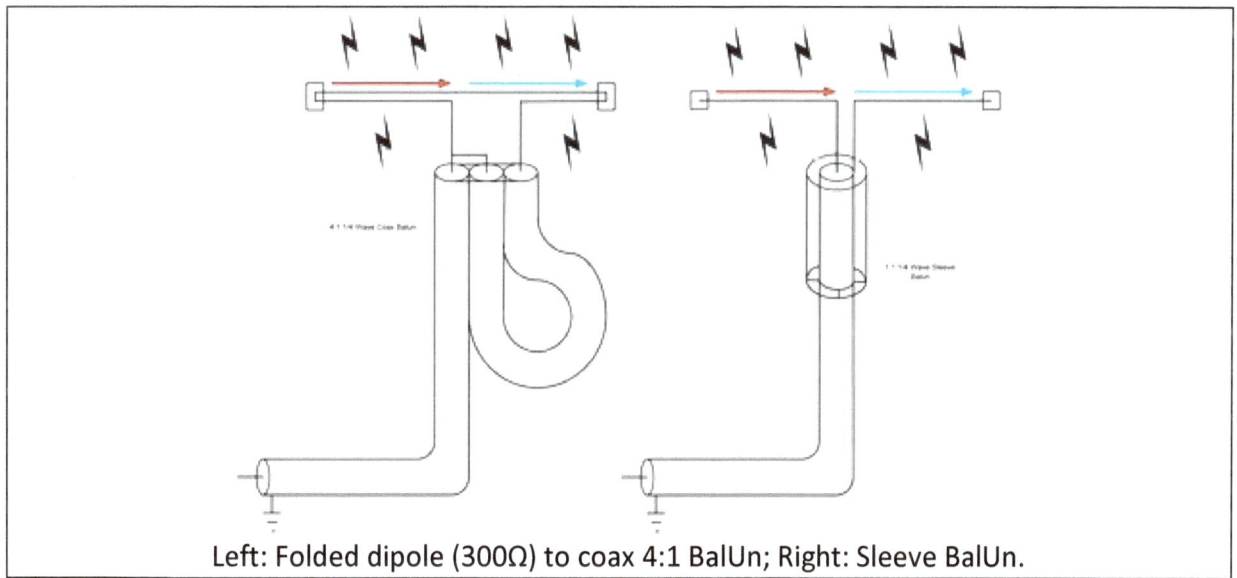

Left: Folded dipole (300Ω) to coax 4:1 BalUn; Right: Sleeve BalUn.

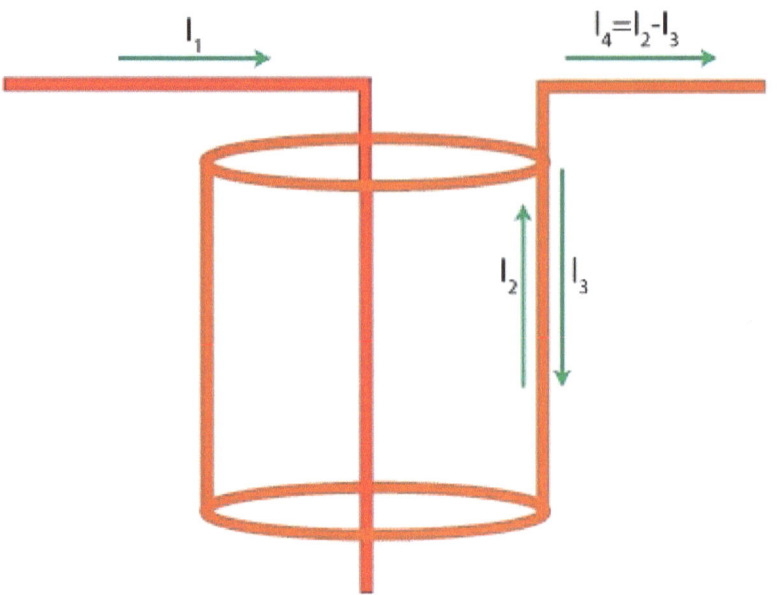

This phenomenon can be described by looking at the following image. Inside coaxial cable there are two currents: I_1 and I_2 with a phase shift by π. Dipole is connected directly to coaxial cable, and the consequence is that one part of I_2 is escaping back through the outer part of the cable's shield, thus I_1 and I_4 are out of balance.

This problem can be also resolved by grounding the outer shield at $\lambda/4$.

The Gamma and Beta (Hairpin) match are not described in this book.

RF Choke

Intro

RF choke is a coil, an inductor, which as all the inductors when put in series, acts as a low-pass filter, i.e. it blocks higher frequencies.

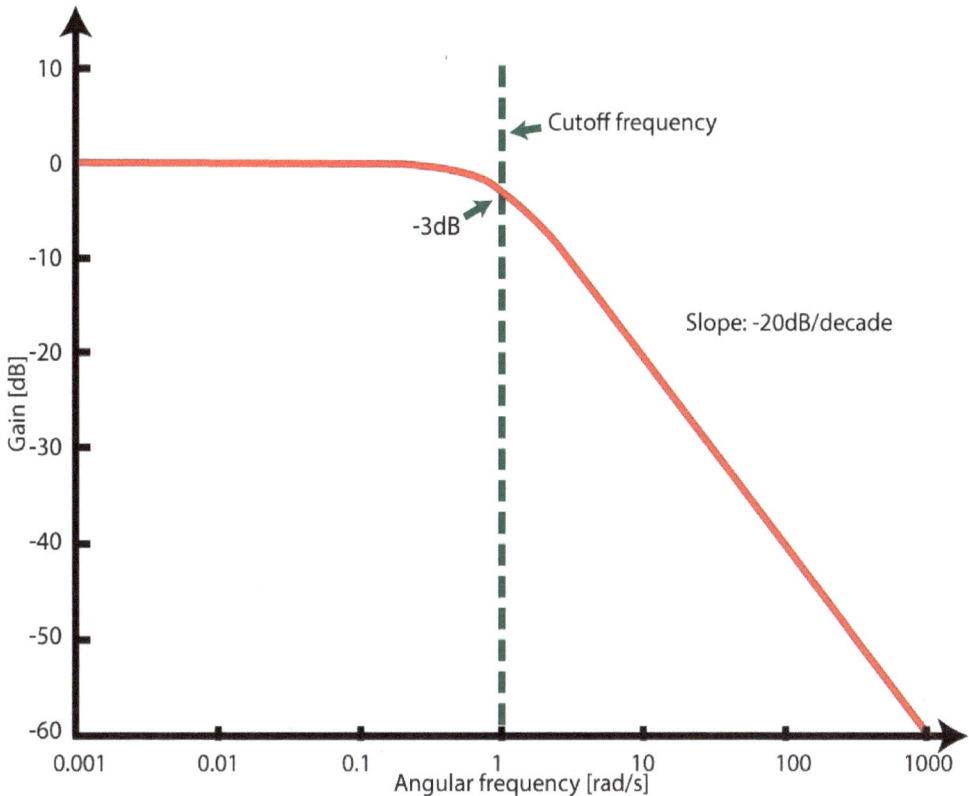

But every coil acts as well as capacitor, thus creating bandwidth filter instead.

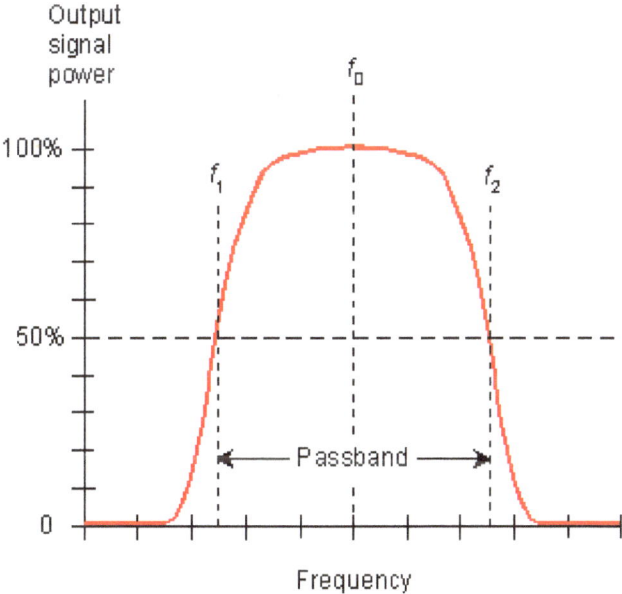

We can see that we must have exact relation of inductance to capacitance for the given frequency.

There are several types of RF choke, but we will look into Ugly Choke BalUn only, of which there are two main types: Air core and Ferrite core. For Radio Amateurs Air core BalUn is much more significant as we do not depend on the core, it is cheaper and even works better for the range of frequencies that are of our interest.

Ugly Choke
Air Core
1:1

Photo of a 1:1 Air Core Ugly Choke BalUn for 27MHz, built from RG-58 coaxial cable, with added SO-239 connector and terminals for Dipole or Inverted V antenna.

Without going into any math, we will give the number of windings of RG-58 cable that need to be wound on the PVC tube with given diameter. Also the length of the cable is calculated and approximate cost of the cable (for Aluminum core and shield RG-58 cable).

Band [m]	Frequency [MHz]	PVC diameter [mm]	Turns	Length [mm]	Cost [$]
80	3.5	110	32	11555.2	9
40	7	110	25	9027.5	7
30	10	110	17	6138.7	5
20	14	110	12	4333.2	3
17	18	110	9	3249.9	2
15	21	110	7	2527.7	2
12	24	110	6	2166.6	2
10	27	110	6	2166.6	2
10	28	110	4	1444.4	1
6	50	50	10	1727	1
3	100	50	8	1381.6	1
2	144	40	9	1271.7	1

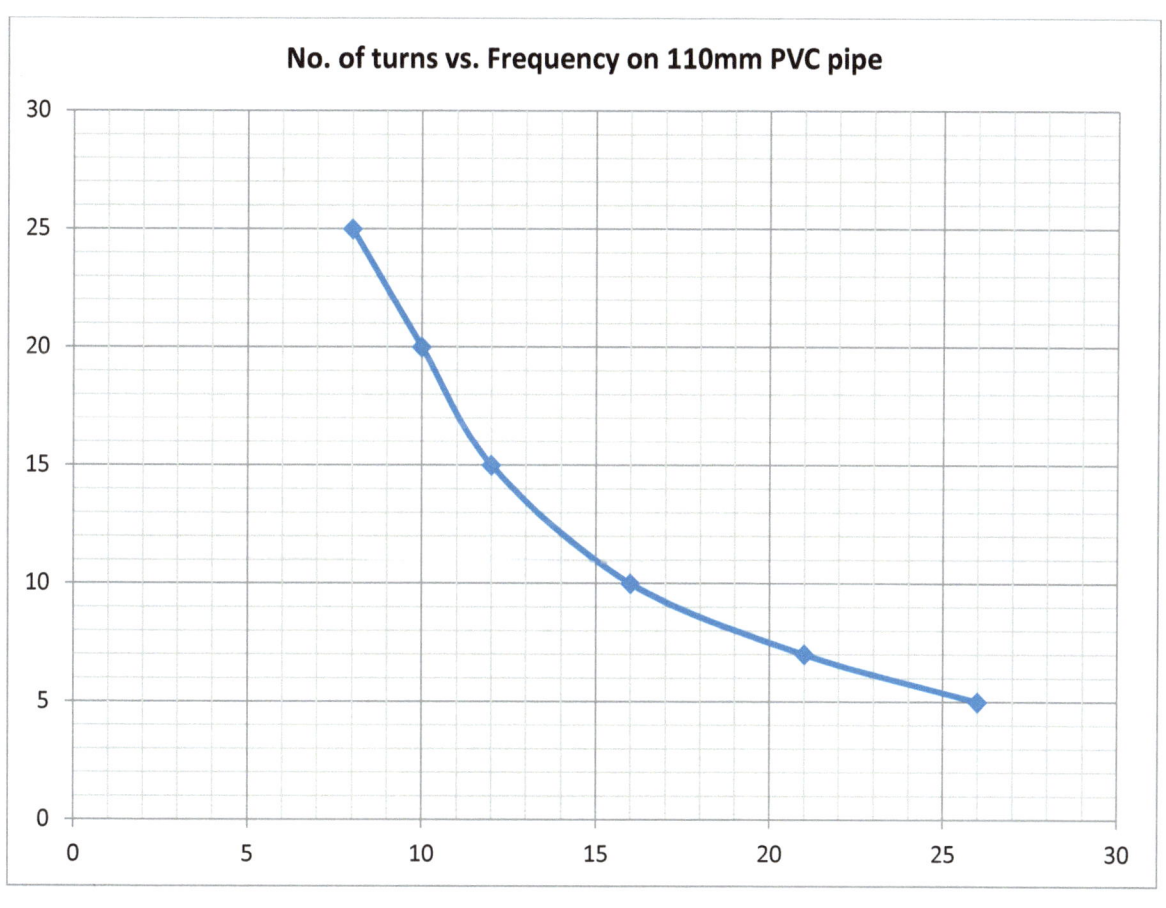

4:1

Two pieces of the same length insulated wire wind to the PVC tube like on image below to get Ugly Choke 4:1 air core BalUn.

For 2m band, enough is 13 turns on the 32mm PVC pipe.

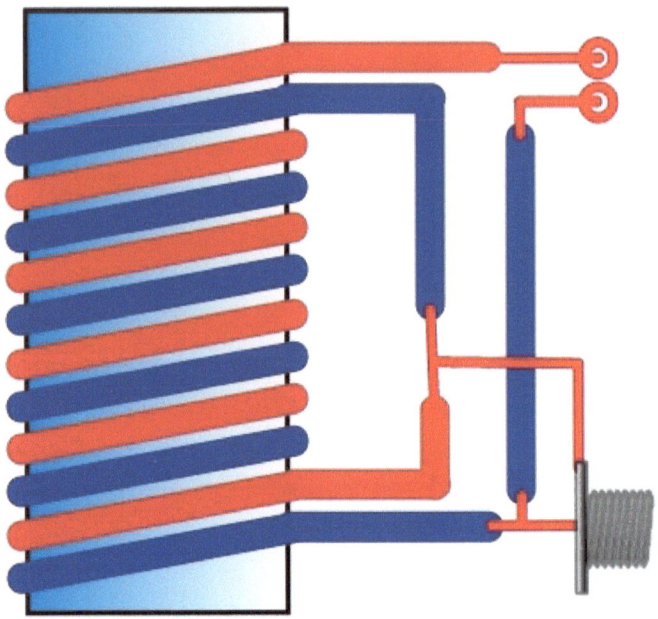

Ferrite core choke 1:1 for 80m band

The principle is the same as for Air core BalUn. For 80m band BalUn will be resonant if RG-58 is wound 24 times on FT240-61 or similar core. It is more compact, but expensive solution, as iron core can cost up to 15 euros.

Delay Line

Coaxial 4:1

U line should be $\lambda/2$ long, with included velocity factor in calculations. Equation for the length of the U line:

$$l = \frac{c}{4f} v$$

v – velocity factor, c – speed of light, f – frequency.

Velocity factor (fraction of the speed of light with which electrons travel inside the cable) of RG-58 and RG-213 is 0.66, RG-6 (with foam as insulator) has 0.81. For other cables refer to the table in the cable section.

The lower the frequency the less accurate length of the U line can be.

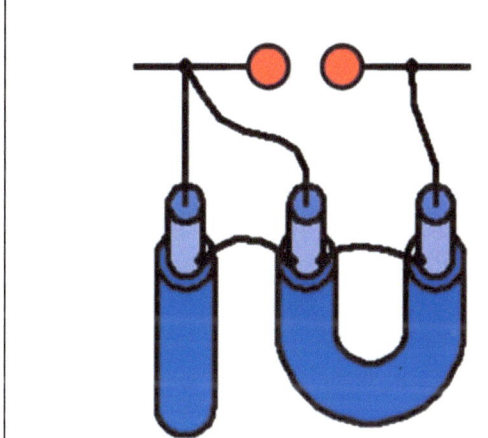

On the top is antenna, right is what we refer as "U line", and on the left is the cable that goes to the device

Frequency [MHz]	Length of U line [mm]
21	4714
24	4125
27	3666
28	3535
50	1980
100	990
144	687
433	228
1280	77
2450	40
5800	17

Parallel cables BalUn

This BalUn consists of two pieces of coaxial cable in parallel with length $\lambda/4$, counting the velocity factor v through the coaxial cable ($v = 0.67$ for the full, $v = 0.82$ for the foam isolation).

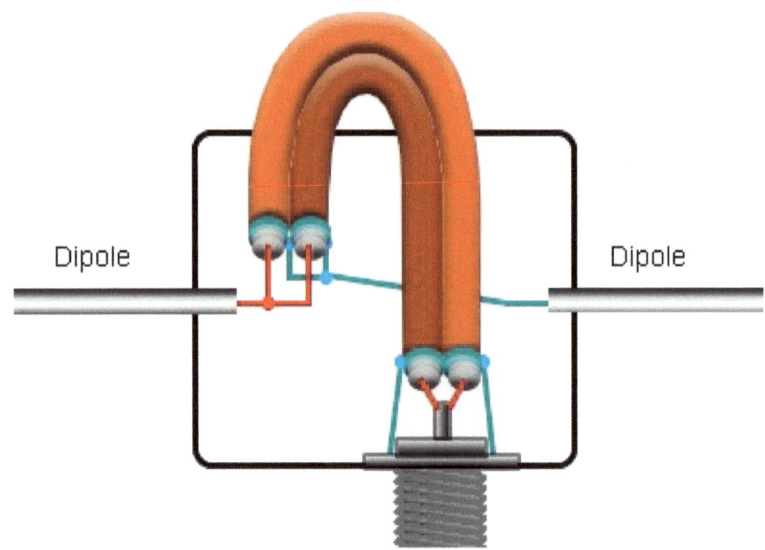

Matching the 50 Ω impedance with the impedance of the antenna:

- 28 Ω with two 75 Ω cables in parallel,
- 12.5 Ω with two 50 Ω cables in parallel,
- 18 Ω with two 93 Ω or with one 50 Ω and one 75 Ω cables in parallel.

Length of the 75 Ω coaxial cable used for the BalUn for matching 28 Ω to 50 Ω:

$$L = \frac{\lambda}{4}\, v$$

Band [m]	Wavelength [m]	Full-PE isolation [m]	Foal-PE isolation [m]
30	29.62	4.94	6.07
20	21.18	3.53	4.34
17	17.53	2.76	3.60
15	14.18	2.36	2.90
12	12.03	2.00	2.47
10	10.60	1.77	2.17
2	2.1	0.35	0.42

Sleeve BalUn

This is very good BalUn and simple to make. The only drawback is that for lower frequencies the pipe needs to be very long, as it should be exactly $\lambda/4$ long. At the beginning, the copper or brass pipe needs to be welded to the braid shield of the coaxial cable, as marked on the image. Thus, it is useful only at higher frequencies.

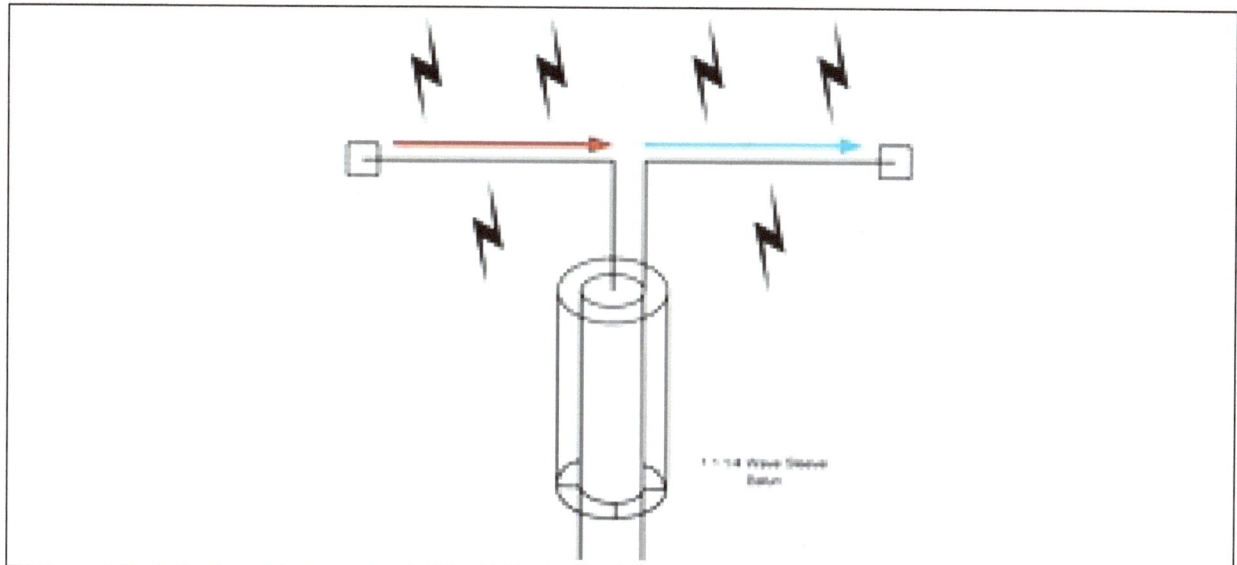

Frequency [MHz]	Length [mm]
433	173
900	83
1290	58
2450	30
5800	13

Torus ferrite core transformer Current-mode BalUn

Advantage of this type of BalUn over the Air Core Ugly BalUn is that it works better at lower frequencies, especially 160m band and it has wider bandwidth. The drawback is the price, since some of the iron cores can be quite expensive, and poor results on higher frequencies. On frequencies higher than 7MHz Air core BalUns are better. When large bandwidth is needed, as with TV antennas, 4:1 BalUn of this type should be used.

First you need to see with the manufacturer that the toroidal core matches the power and the frequency.

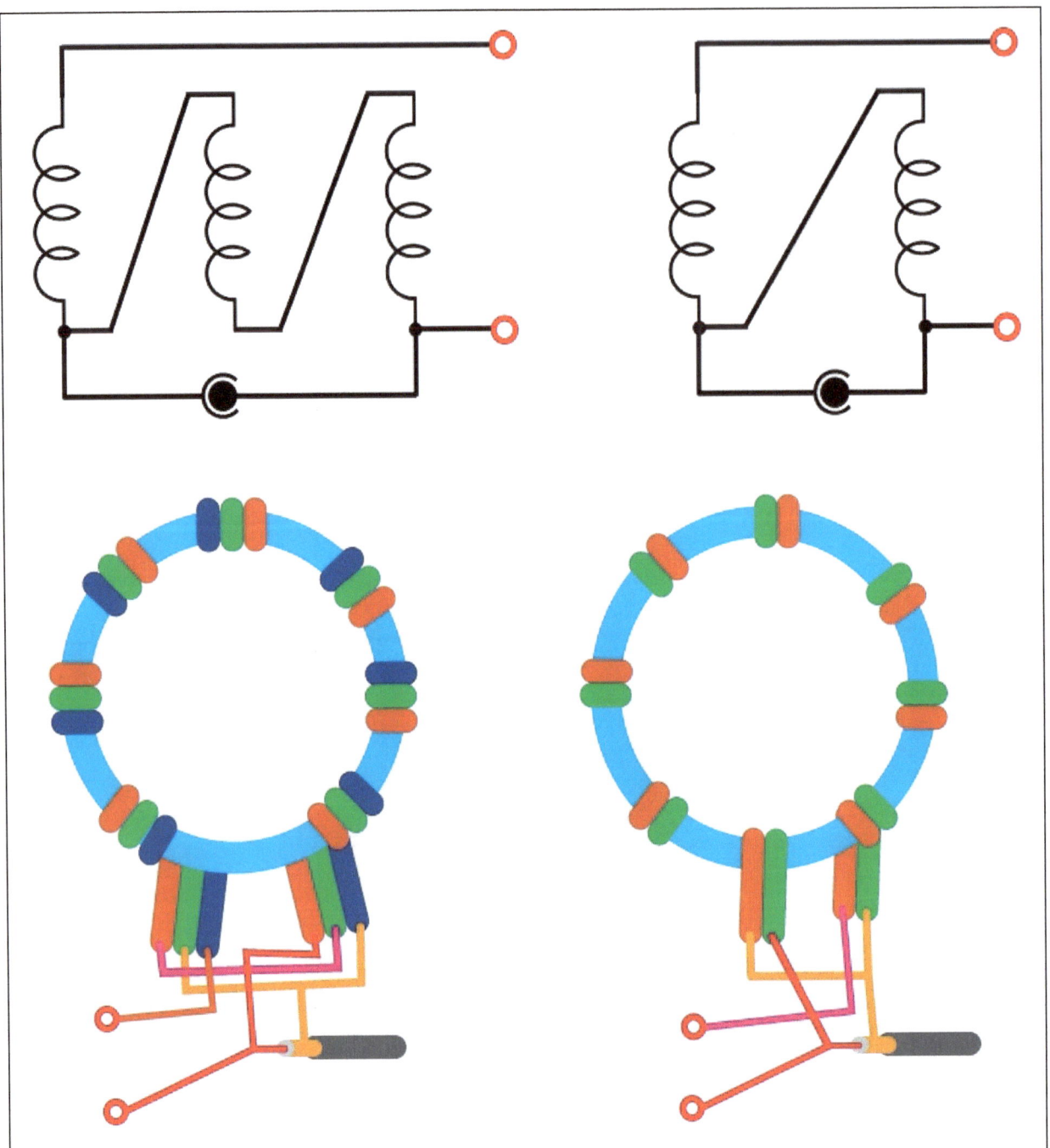

1:1 BalUn on the left and 4:1 on the right. Each inductor shown above has the same number of turns, and that number is given in the table below. Each coil should be wound on the same ferrite toroid core.

Enameled copper wire is used for threads. Wire diameter should be chosen so that windings fit the core without overlapping. Windings should be tight.

Core type up to 30MHz	Power [W]	Turns
T-80-2	60	25
T-106-2	100	16
T-130-2	150	18
T-157-2	250	16
T-200-2	400	17
T-200A-2	400	13
T400-2	1000	14

Cables

The most common cables used are 50Ω, 75Ω or 300Ω. This is because they provide a good match to antenna or device you attach them to. Without going into theory to show how much proper impedance is important and to talk about SWR (standing wave ratio), we will instead just list the types of cables, when they are best used and how much they attenuate at which frequency. But beware, cable type defines only its geometry and impedance, not the materials it is made from. You will likely see three types (also there are different types of insulation, with different velocity factors):

1. Aluminum shield and aluminum hot wire,
2. Aluminum shield and copper hot wire,
3. Copper shield and copper hot wire.

This affects price as well, and proper choosing of the cable depends on application. For example, any cable with aluminum should not be used for Wi-Fi. There is also a problem with welding the connector, unless crimping is being used.

Some earlier cables did not use aluminum or copper foil. Although they have much richer shield than the new cables, their attenuation can be greater. Avoid using old cables when possible, because they tend to break or their insulation rots, and you will be left wondering why your antenna is not working.

Some of those cables come with additional steel wire, so that they can hang from building to building. Otherwise snow or some bird could stretch them. Some cables are for internal or external use. Those meant for internal use cannot withstand rain, snow or sun – their insulation will start to crack and eventually the whole cable will break.

All described cables are coaxial, but twin-line.

300Ω is used with FM/AM. 75Ω is used in FM, TV and some cheap Wi-Fi. 50Ω is used for everything else, but it is the most expensive type.

75Ω

RG-6

This is the most common cable for analogue or cable TV. It has good price to attenuation ratio. It is also being used for FM tuners or even for some cheap Wi-Fi antennas (although this is quite a poor choice and those antennas should be avoided). For FM radio it should be used if on the back of FM tuner, it is stated 75Ω unbalanced for the antenna. It is 8mm thick.

RG-6 cable with copper hot wire and aluminum shield

RG-11

This is a thicker (10mm) 75Ω cable than RG-6, with extra shield. It has less attenuation and is generally better cable.

RG-59

The cheapest variant of the three. It is commonly used for indoor antennas and it should be avoided for other purposes.

300Ω

Twin-lead

Twin-lead cables come with many impedances. Although the most common is 300Ω, one can see 450Ω as well. 300Ω twin-lead has 20-gauge wires 7.5mm apart. This type is important because it is commonly used with some types of FM tuners, and unlike coaxial cables is balanced.

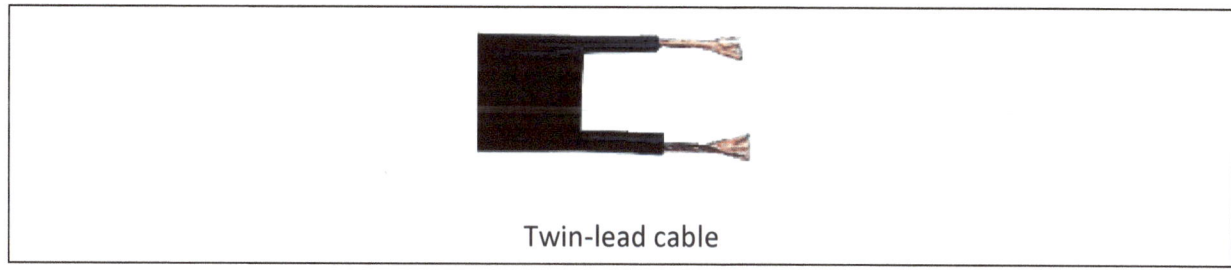

Twin-lead cable

50Ω

RG-58

It is often used for low-power signal. There are lot of variants of this cable in the means of core material, shield material, shield coverage, solid core or braided wire and insulation. Outer diameter is 5mm. Plain RG-58 cable has solid core, while RG-58U has braided wire, which is better at high frequencies. The prices of different variants of RG-58 can vary up to four times.

This is the most commonly used cable for cheap Wi-Fi or among HAMs.

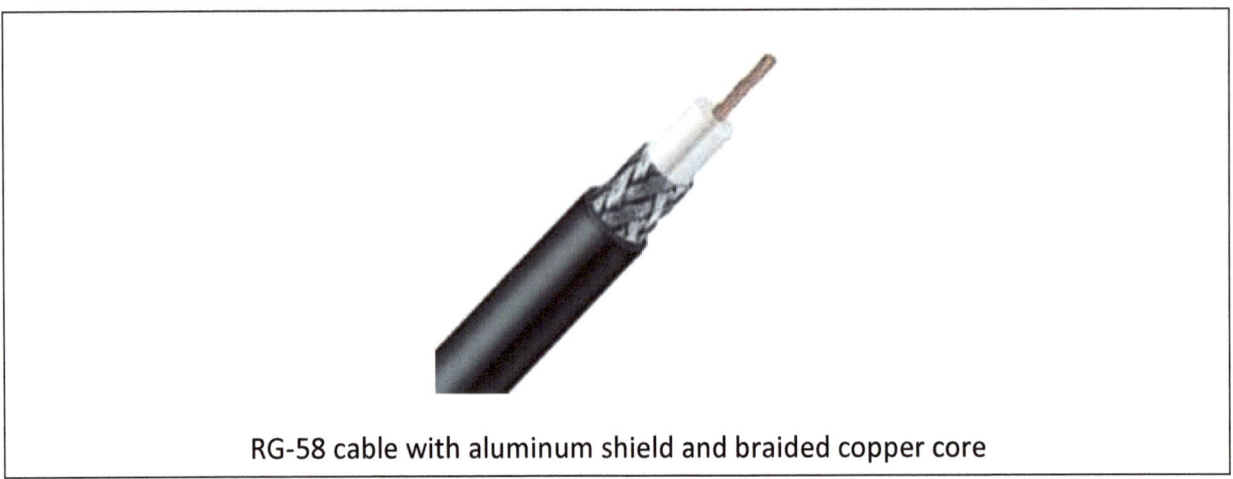

RG-58 cable with aluminum shield and braided copper core

H155 or RF-50

Most common name is H-155, but some manufacturers label it as RF-50. It is similar to RG-58, but is slightly thicker and has better performances i.e. less attenuation at a slightly higher cost, making it a best buy for the Wi-Fi cables for clients.

RG-213, LMR-200, LMR-400

RG-213 is over 10mm thick, the same as RG-11. In the means of attenuation, it comes between two other common types, LMR-200 and LMR-400, latter being the best and the most expensive cable. All three types are quite common and are used when high power or good signal quality is needed.

RG-213 cable with copper shield and braided copper core

Coaxial Cable Comparison chart

Cable	Z [Ω]	Velocity Factor	Attenuation at 750MHz [db/100m]	[db/100ft]
RG-6	75	0.75	18.54	5.65
RG-11	75	0.66	11.98	3.65
RG-59	75	0.66	31.82	9.70
RG-58	50	0.66	42.98	13.10
H155	50	0.79		
RG-213	50	0.66	19.55	5.96
LMR-200	50	0.83	29.53	9.00
LMR-400	50	0.85	11.61	3.54

Connectors

There are hundreds of types of connectors, however there is some standardization, so that only few are being used for the topic we are researching, and yet even less for the specific branch. Connectors we will be describing are:

- PL/SO-239
- BNC
- N
- RP-SMA
- RP-TNC
- RF or Belling-Lee or IEC 169-2 or TV or PAL
- F

Same type connectors can have different impedances. The impedance should match cable's impedance.

Connector types
PL-259/SO-239

PL-259 is male connector that is labeled SO-239 for its female variant. Its other names are UHF connector and Amphenol coaxial connector. It is popular with Radio Amateurs, as it can withstand powers of over 2kW. There are versions for different cables, mainly for RG-58 or RG-213. They are 50Ω only.

PL on the left and SO-239 on the right.

BNC

BNC can be commonly found on radio stations. They can withstand less power than PL connectors, but are better at higher frequencies, all up to 2GHz. They are standard connectors on Oscillators. There are 50Ω and 75Ω variants, and usually for thinner cables, like RG-58. There are also variant for crimping, screwing or welding it to the cable.

They can be used as a cheap replacement for RP-SMA or N connectors for Wi-Fi 2.4GHz, since they are much better at higher frequencies than F connectors.

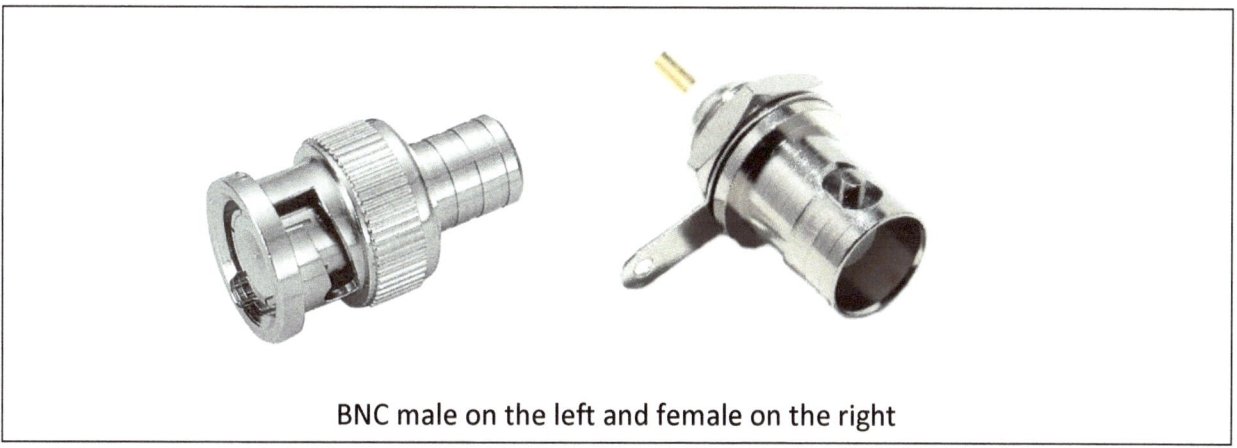

BNC male on the left and female on the right

RP-TNC

This connector is listed here only because Cisco uses them for their Wi-Fi routers. Unfortunate is that they are twice as expensive as RP-SMA and 50% more expensive than N connectors. They work up to 11GHz.

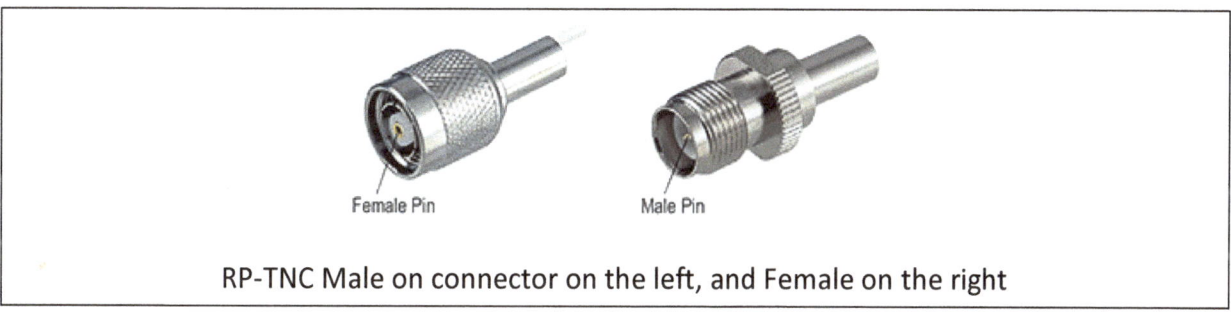

Female Pin   Male Pin

RP-TNC Male on connector on the left, and Female on the right

F

F connector is most commonly used for lower frequencies and can be found on TV Satellite Receivers, Cable TV or Cable modems. It is the cheapest of the all the connections mentioned. There are variants for 50Ω that can be mounted on RG-58 cable, and 75Ω variant that is suitable for RG-6.

F connector uses cable's core wire as the pin of the connector. RG-6 cable has solid conductor core wire and it is easily mounted with screwing. 50Ω variant is usually for crimping, and since it has multiple thin wires for a core, it must be welded to be used as a pin.

It can be used for cheap Wi-Fi at 2.4GHz, but it is usually a poor choice since there is both loss in dB and signal quality.

Female F to Female F is the best option when it is needed to extend RG-6 cable.

Left to right: Male F connector for RG-6, Male F connector for crimping, Female F to Female F cable extender

RF or Belling-Lee or IEC 169-2 or TV or PAL

These are all the names used for the same type of connector. To raise the confusion, the name RF can also be used for PL connectors. Here we will refer to it as the RF connector. Male RF connector is being used on TV antennas (female is inside TV). With some FM tuner receivers, the opposite is true, female is used on a cable while male is in the device. It is very easy to mount it, but it gives the worst performances of all the connectors mentioned. Variants of it are with metal or plastic housing. Since they are cheap, do not save by buying the one with the plastic housing as it can cause your TV or FM signal to be noticeably poorer than it should be.

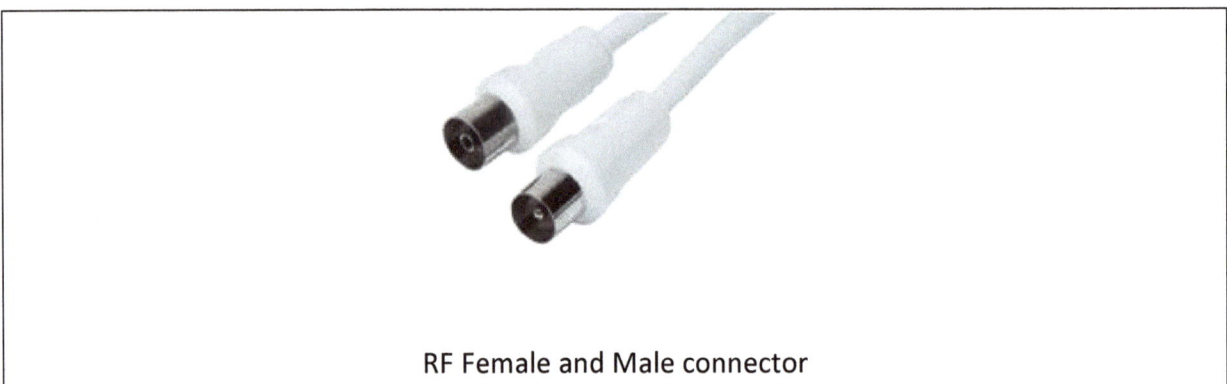

RF Female and Male connector

RP-SMA

RP stands for Reverse-Polarity. These connectors are most commonly used for Wi-Fi, to connect AP/Router or PCI/USB Wi-Fi card with 50Ω cables. On the side with the antenna it is far more common to find N connector. They are 50Ω only, but variants can be in materials it is made from, crimp or screw-on, for RG-58 or RG-213 cable. All but Cisco (Linksys) devices use it. It is very good in terms of quality of signal, but can withstand less power than N connector.

Male connector type is usually on cables, while female is most often found on devices.

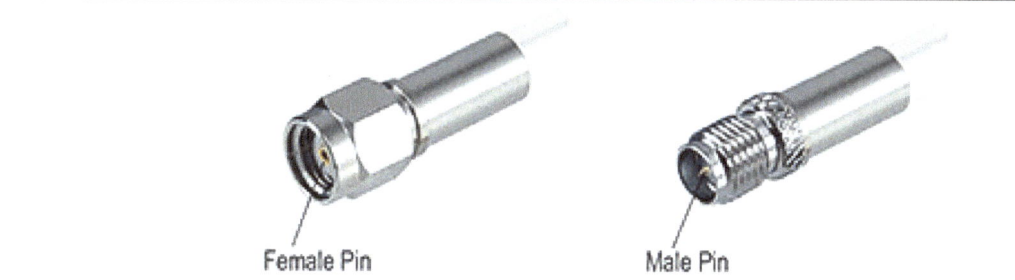

Female Pin   Male Pin

On the left is RP-SMA Male, which has Female pin, on the right is Female type with male pin

N

It is the most common connector on Wi-Fi antennas and with Radio Amateurs, mainly at UHF range. It is rated to transmit up to 18GHz, and up to 5kW at 20MHz or 500W at 2GHz. Its only drawback can be its price.

Although there are 50Ω and 75Ω variants, 75Ω are only used for Cable TV systems. In terms of mounting a connector, it can be crimp or screw-on if for the cable, if it is for device, it can come with four holes or a thread and a screw, same as the SO-239 connector variants. Female is most often used on devices or antennas, while the male is most often found on cables.

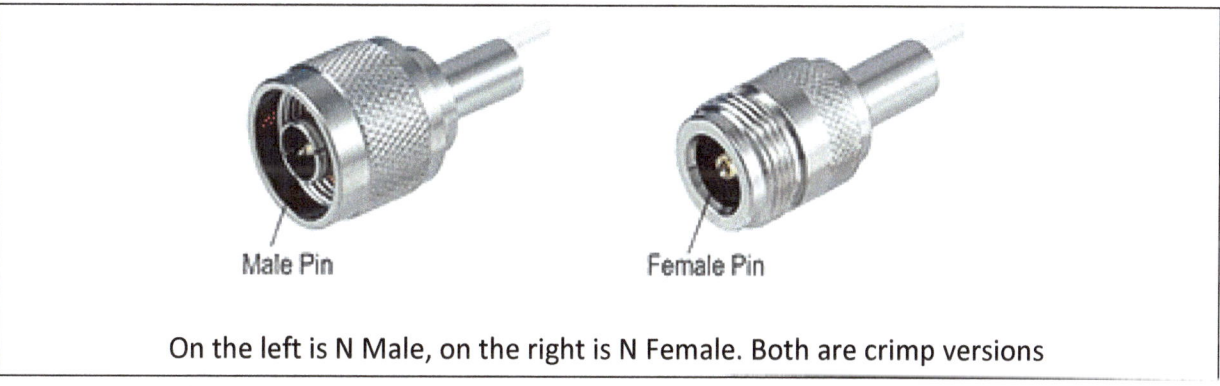

Male Pin   Female Pin

On the left is N Male, on the right is N Female. Both are crimp versions